**Between Text, Meaning and Legal Languages**

# Foundations in Language and Law

**Editors**
Janet Giltrow
Dieter Stein

**Volume 8**

# Between Text, Meaning and Legal Languages

Linguistic Approaches
to Legal Interpretation

Edited by
Jan Engberg

**DE GRUYTER**
MOUTON

ISBN 978-3-11-221478-7
e-ISBN (PDF) 978-3-11-079965-1
e-ISBN (EPUB) 978-3-11-079969-9
ISSN 2627-3950

**Library of Congress Control Number: 2023943633**

**Bibliographic information published by the Deutsche Nationalbibliothek**
The Deutsche Nationalbibliothek lists this publication in the Deutsche Nationalbibliografie;
detailed bibliographic data are available on the internet at http://dnb.dnb.de.

Chapter "Natural semantic (legal?) metalanguage. What can legal theory learn from Anna
Wierzbicka?" © Mateusz Zeifert

© 2025 Walter de Gruyter GmbH, Berlin/Boston
This volume is text- and page-identical with the hardback published in 2024.
Cover image: kokouu/E+/Getty Images
Typesetting: Integra Software Services Pvt. Ltd.
Printing and binding: CPI books GmbH, Leck

www.degruyter.com

# Contents

Jan Engberg
**Introduction —— 1**

## Section 1: **Investigating legal interpretation and argumentation**

Mathilde Barraband, Anne-Marie Duquette, Julien Lefort-Favreau
**The Dulac affair and the triple game of contemporary art —— 9**

Stanisław Goźdź-Roszkowski
**Argumentation, rhetoric and legal justification. The case of Poland's Constitutional Tribunal ruling on abortion —— 29**

Giovanni Tuzet
**The pragmatics of evidence discourse: Ostensive acts —— 47**

## Section 2: **Looking at language to investigate legal challenges**

Daniel Greineder
**Illusions of a common Language: Impressions of an arbitration practitioner —— 67**

Jacqueline Visconti
**Pragmatic features of Italian court proceedings —— 79**

Jakub Eryk Marszalenko
**Politeness Matters: What honorifics can tell us about accuracy in Japanese-English court interpreting —— 91**

Caroline Laske
**Textual representation as a conceptual tool: Big data analysis of legal language —— 117**

## Section 3: Theories of sense and meaning for legal investigations

Weronika Dzięgielewska and Wojciech Rzepiński
**What is practical about law? Contemporary legal philosophy on legal practice —— 145**

Mateusz Zeifert
**Natural semantic (legal?) metalanguage. What can legal theory learn from Anna Wierzbicka? —— 173**

**Index —— 205**

Jan Engberg
# Introduction

Legal interpretation in the wide sense intended in this work is about analysing and establishing the meaning of language, i.e., about understanding expressions in texts. Hence, the basic practical problem to be solved is to find out what a specific text may or must mean. In principle, this is not difficult, it is what we do all the time in communicative exchanges with others in our daily lives. However, the specific communicative and pragmatic characteristics of legal communication generate specific challenges. For one thing, legal texts appear in the medium of writing, which means that the panacea of communicative understanding, i.e., explicit and implicit cues from direct interaction are not available. Interpreters must make do with hypothetical interpretations from the point of view of aspects of the texts that the interpreter sees as important, based on how the interpreter sees the context. Secondly, legal texts are aimed at the communicatively difficult construct of general public, i.e., an unspecified audience. As modern text linguistic has demonstrated, there is no such thing as a non-contextual meaning of concrete texts. Hence, as textual meaning depends upon understanders and contexts, in legal interpretation it is necessary to make assumptions about the characteristics of the intended understander.

As an answer to this challenge, most jurisdictions have developed a set of accepted methods to be applied in statutory interpretation, probably based on a systematisation of everyday principles of understanding. The rationale behind such standardisation is to supply the legal expert with a set of agreed tools for the process in order to avoid a too subjective interpretation. For instance, in Germany the following four methods are applied:
- Grammatical interpretation, starting from the wording of the statutory provisions to be interpreted.
- Systematic interpretation, starting from the position of the statutory provisions in the relevant legal system and the relations to related provisions.
- Historical interpretation, starting from the intentions and motives of the legislator as they are accessible in historical sources.
- Teleological interpretation, starting from the idea that the provision should be perceived as a just solution to the problem it wants to solve.

Such sets of interpretation methods are valuable guidelines for legal practitioners and enable them to express quite clearly and systematically on what grounds an

**Jan Engberg,** Aarhus University

https://doi.org/10.1515/9783110799651-001

interpretation has been carried out. They are also easy to understand as they reflect basic criteria involved in understanding what others say in communicative settings (what exactly did the other person say, how does it match what I know about the background of what we talk about, how does it match what was said previously, how does it match the goals at hand in the situation?). Problematic is that choosing different approaches may very well yield different results. So, legal interpreters may tell us what the principle for their chosen understanding is. But solely on the basis of the methods shown above, they may not tell us why they choose a specific method. Consequently, they may not tell us exactly how they came from a specific formulation of a provision to the actual interpretation in a fully consistent manner.

One of the challenges in such situations is that the road from utterance to meaning in the mind is not reflected upon to any great extent in the methods. Instead, the 'mechanism' is seen as a black box. As modern text linguistics, cognitive linguistics and pragmatics have a lot to say about such processes, statutory interpretation as well as argumentation has been one of the central objects of study in the development of legal linguistics. The discipline constitutes the fruitful coming together of interests from several disciplines and a good forum for interdisciplinary cooperation aimed at assessing what is going on in processes of structured understanding in the field of law. The present book wants to present a variety of approaches all interested in law, language, and generation of meaning and sense. The overall common interest is in how theoretical approaches linked to linguistic studies in a broad sense may help us understand better legal meaning and legal practice. The selection of contributions to this volume demonstrate how a cross-disciplinary cooperation like the one driving legal linguistics may help improve the actual insights in the studied object, mainly through enlightening the black boxes and blind spots of the respective disciplines combined.

The chapters are divided into three sections. The first section consists of works investigating aspects of actual legal interpretation and argumentation, i.e., looking at how meanings are found in specific legal settings. Investigating actual legal argumentation, the first chapter *The Dulac affair and the triple game of contemporary art* by Mathilde Barraband, Anne-Marie Duquette, and Julien Lefort-Favreau presents a case study of the argumentation in a court case involving the interpretation not only of legal concepts, but also of the concept of sociology of art. In the context of a course in art performance at a Canadian art school, a student presented a project for a student exhibition evoking a child kidnapping as well as acts of violence against children. The chapter investigates the lines of argumentation in the two courts that decide the case in the first and the second instance. The first court sees the accused as guilty, whereas he is acquitted in the second court. The case is as an exemplary embodiment of a conflict of norms. It may be understood as an effect of the autonomization of art, in particular the autonomization of its values, its definitions of beauty and goodness, regarding common sense and to the legal norm.

The second chapter investigating actual legal argumentation is by Stanisław Goźdź-Roszkowski (*Argumentation, Rhetoric and Legal Justification. The case of Poland's Constitutional Tribunal Ruling on Abortion*). The chapter studies a recent politically controversial ruling by Poland's Constitutional Tribunal on the banning of abortion. Point of departure is legal argumentation theory, incorporating certain elements of classical legal rhetoric (*ethos*, audience effect, *topoi* and *kairos*). In the chapter, the author reconstructs the judicial reasoning by using weighing and balancing as a principled method for external justification. Based on the analysis, the author establishes that although a wide range of argumentative means are invested in the decision it cannot be considered as effective and acceptable because of the court's undermined legitimacy and a failure to convince key audiences. In other words, the unusual argumentation undermines the power of the argumentation.

The third chapter, by Giovanni Tuzet, with the title *The Pragmatics of Evidence Discourse: Ostensive Acts* investigates a specific instance of legal interpretation: the interpretation of something as a juridical proof. This process of juridical proof requires some ostensive act, which means that it must present and declare something as evidence to prove a claim. Building on previous work on the pragmatics of evidence discourse, the author especially focuses on the dynamics of ostensive acts. The chapter intends to clarify how such acts work in the context of legal fact-finding, focusing particularly on the use of indexical words. It concludes by claiming that ostensive acts are necessary but not sufficient to legal fact-finding, for ostension must be followed by argument.

The second section is constituted by works that investigate linguistic aspects in order to throw light on legal problems. The chapter *Illusions of a Common Language: Impressions of an Arbitration Practitioner* by Daniel Greineder is interested in the specific characteristics of international commercial arbitration to be a jurisdiction without a fixed language. Parties may freely choose the procedural language they can agree upon. What they often take is English, which is generally used as lingua franca in the context of international business. Based on his long-standing experience in international commercial arbitration, the author describes potential problems emerging from a lack of awareness of the influence of language choice. Especially, the author develops the idea that English as a widespread, but by no means universal, language of arbitration has contributed to the development of international arbitral practices and strengthened it institutionally. However, users of English who do not have it as their primary (working) language may lack a high level of proficiency in English, and the development of English as a language of arbitration has not kept up with the internationalization of arbitration itself, either. As a result, a form of legal English has emerged that is often unsubtle and deficient, and thus a victim of the success of arbitration.

The second chapter in this section is written by Jacqueline Visconti and has the title *Pragmatic features of Italian court proceedings*. In the chapter, she presents first results of an interdisciplinary project on clarity and understandability in counsel documents in Italian courts. The first part of the presentation describes an innovative tool for anonymising documents, which due to its efficiency is a very central tool in convincing practitioners of making authentic documents available for research, without preventing relevant corpus linguistic research. In the second part, first qualitative studies of the genres in the corpus are presented, demonstrating basic characteristics of the genre, among other things from the point of view of macrostructure.

In the third chapter of this section, *Politeness Matters: What honorifics can tell us about accuracy in Japanese-English court interpreting*, Jakub Marszalenko presents the problems that especially legal interpreters working between Japanese and English in court settings encounter due to the basic differences between the honorific systems in the two languages. The main problem is the very complex system referred to as *keigo*, which governs expression of respect between people of different age, profession, or other type of social standing. The system is visible in lexis, inflection, and pragmatics, but particularly prominent in verbs, which in spoken Japanese take either casual or formal inflection forms. The Japanese system is not (easily) translatable into English, so there is going to be a loss of meaning in this direction, whereas the interpreter may have to insert meaning when interpreting from English to Japanese. These characteristics challenge the traditional formulations of the ideal of accuracy in translation in Japanese court. The investigation is based on empirical work and offers examples of different types of solutions to the problems.

In the last chapter of this section (*Textual representation as a conceptual tool: Big data analysis of legal language*), Caroline Laske argues for a sociolinguistic approach to legal linguistics. Such an approach should supplement the traditional approach focusing upon the specialized aspects of legal language like vagueness and terminological precision often subsumed under the title of LSP (Languages for Specific Purposes). The traditional approach is characterised by specialized performativity and carrying out the characteristic communicative and legal tasks of law. The proposed sociolinguistic approach, on the other hand, is oriented towards law as a tool of social governance and interested in describing a society's perceptions of its identity, including the hegemony of its structures, its moral, social, geo-political and cultural constructs, and its economic needs, but still through the study of linguistic structures. It thus starts from the characteristic of language that it reflects the social factors in the context in which it is used. As a specifically apt tool for investigating such questions, she proposes the use of corpus linguistics on large corpora for investigating ordinary meanings in legal language, semantic objectivity as a result of linguistic convention, and the textual representation of specific categories of persons (in this case women in legislation in Normandy, Saxony, and Norway in the thirteenth century).

Finally, the third section contains two chapters that propose linguistic theories of understanding as methods for solving problems of legal practice as well as legal theory. In their work with the title *What is Practical About Law? Contemporary Legal Philosophy on Legal Practice*, Weronika Dzięgielewska and Wojciech Rzepiński present a descriptive approach from language philosophy (Robert B. Brandom's concept of meaning-use-relations aimed at explaining how linguistic meaning is generated from language use). This approach is proposed for describing meta-theoretically and comparing how the concept of practice is applied (= understood) in four different approaches to legal philosophy. Hence, the link to language and linguistics in this work is constituted by applying a linguistic meaning theory connected to practice for describing differences in the practice concepts in approaches to legal philosophy. The authors see the investigated theories from legal philosophy as languages and thus look at how these 'languages' use the concept of practice. The four approaches compared are Harts *The Concept of Law*, Dworkins *Law's Empire*, Raz' *Investigations on the Nature of Practices Underlying Law and Value*, and Pavlakos' *Practice Theory of Law*.

In the last chapter of the book (*Natural Semantic (Legal?) Metalanguage. What Can Legal Theory Learn from Anna Wierzbicka?*), Mateusz Zeifert focuses upon a specific linguistic semantic theory, viz. the theory of Natural Semantic Metalanguage (NSM). The basic claim of the theory is that it is possible to establish a list of semantic primes, i.e., concepts that are claimed to be primary (= they cannot be explained in simpler terms) and universal (= they are lexicalised in all human languages). The primes are related to general human conceptualisation of tasks and (bodily) experiences and for that reason proposed to be universal. NSM is used for describing meaning in different languages in a comparable format. Such descriptions could be used for enhancing the comprehensibility of legal texts (by being guided by the semantic primes). It may also be used for semantic analyses of legal terms, which typically lack coherent methodology. Due to its proposed universality, Natural Semantic Metalanguage is well-suited for comparing the meaning of words across languages. For this reason, descriptions from the approach are relevant for comparative law as well as for the drafting of multilingual and culture-neutral documents in settings of international law.

The eight chapters in this volume constitute an attempt to give an overview to some of the many facets of legal linguistics. It is the editor's hope that the volume may function as inspiration for the search for approaches from especially (text) linguistics, pragmatics, and cognitive semantics that may be instrumental in throwing light on black-box aspects of law and legal interpretation and argumentation.

# Section 1: Investigating legal interpretation and argumentation

Mathilde Barraband, Anne-Marie Duquette, Julien Lefort-Favreau
# The Dulac affair and the triple game of contemporary art

**Abstract:** The Dulac Affair begins in the spring of 2013. David Dulac is then a third-year student at Laval University's School of Art in Quebec City (Canada) and is submitting a project for a student exhibition. Dulac's proposal is a performance project that evokes a child kidnapping as well as acts of violence against children. The document goes to the director of the art school, then to the school's psychological support team, who warns the school's security. The school security believes there is a threat and notifies the police; on March 28, 2013, Dulac appeared in court for "transmitting or causing threats to cause death or bodily harm to children in the region's elementary schools". It is from a sociology of art perspective that we have chosen to analyze this case. More precisely, it is as an exemplary embodiment of a conflict of norms that we want to highlight. This case, like many others, can be understood as an effect of the autonomization of art, in particular the autonomization of its values, its definitions of beauty and goodness, with regard to common sense and to the legal norm. But it is also a conflict between divergent systems of evaluation within the art field itself.

**Keywords:** Performance Art, Art School, Autonomization, Conflicts of Norms, Nathalie Heinich

## 1 Introduction

What we call "the Dulac affair" began in spring 2013 when David Dulac, a third-year art school student at Laval University, submitted a project for an exposition scheduled for May of that year. The aim of this exposition was to showcase final year student's artworks to the general public. It was organized by a committee of peers. Another graduate, referred to here as student A., was tasked to collect her colleagues' proposals, submitted on a voluntary basis, and hand them to the exposition organizer. Upon opening the folder submitted by David Dulac, she read the following handwritten notes on a lined sheet of paper:

---

**Mathilde Barraband, Anne-Marie Duquette,** Université du Québec à Trois-Rivières
**Julien Lefort-Favreau,** Queen's University

> Description of David Dulac's project for the graduate exposition
> I don't have an outline of the project at the moment, but I'll describe it in general. My project will be performative and will consist of first, kidnapping as many child (sic) as possible by using candy, video games or iPods to lure them into my car near an elementary school in the region and then tying them up in old potato sacks or straws (sic) sacks; next, during a performance, once they will (sic) all hanging (sic) from the ceiling, I'll put on a blindfold and hitting (sic) them with a heavy object. The point of the work is to demonstrate how sweet little innocent children will age throughout the contemporary world into the spineless adults of the future. I of course will represent humanity or its legacy; that depends on point of view.[1]

Student A would later describe her first reaction as follows: "In my opinion, this isn't feasible within the context of the exposition, besides that [. . .] it borders on disturbing."[2] She then decided to take the document to the school director, who notified psychology support services, who, in turn, notified the university security services. The latter deemed it a threat and called the police. On March 28, 2013, Dulac appeared in court for "conveying or causing the reception of threats of death or bodily harm to children in the region's primary schools."[3]

Unusual for a first-time offender, Dulac was placed in temporary detention for the four months between his court appearance and trial in the Criminal and Penal Division. The crown prosecutor based his argument on the escalating gravity of Dulac's artistic projects. He asked several professors and students who knew Dulac to testify to the young man's work prior to the project submitted for

---

[1] "Description du projet pour l'expo des finissant (sic) de David Dulac / Je n'ai pas d'image à fournir du projet pour le moment je vais décrire en gros ce que je vais présenter. Mon projet sera performatif et consistera d'abord à kidnapper le plus d'enfant (sic) possible en les attirant dans ma voiture près d'une école primaire de la région à l'aide de bonbon, de jeu vidéo ou de gadget, style iPod, et de les enfermer dans des vieilles poches de patates ou de sacs de pailles (sic), et pendant une performance, une fois qu'ils serons (sic) tous accroché (sic) au plafond, je me banderai les yeux je les frapper (sic) avec une masse de fer. Le sens de l'œuvre sera de démontrer comment les beaux et petits enfants innocents vont vieillir au travers le monde contemporain pour devenir des adultes amorphes de demain. Moi je représenterai bien sûr l'humanité ou son héritage, cela dépend du point de vue." The text is reproduced here exactly as in the judgment. A [sic] was not added in places that called for one; only those highlighted by the Court were left in. See *The Queen v. David Dulac* [judgment], Court of Québec, July 19, 2013, § 7. All following quotes and citations in this paper are freely translated from the original French.
[2] "selon moi, c'est pas réalisable dans le cadre de l'exposition pis [. . .] c'est limite inquiétant", *The Queen v. David Dulac* [trial transcription], Court of Québec, July 11, 2013, 134. To save space, all references in the text to the two transcriptions of the Court's examinations appear as (*TD1*) for the July 11 transcription and (*TD2*) for the July 12 transcription, followed by the page number.
[3] "transmis ou fait recevoir des menaces de causer la mort ou des lésions corporelles à des enfants des écoles primaires de la région." Judgment, *The Queen v. David Dulac*, Court of Québec, July 19, 2013, § 1.

the graduating student's exposition. The defense did likewise. The result was a description of an entire series of Dulac's artistic projects, including one work involving photographs of fermented sperm for a painting technique course and another adapting the concept of the video game *Duck Hunt*[4] to the context of a school shooting, where the player was required to shoot "bandits" to save students (*TD1*, 28). But what the court found most shocking was two performances held in the "Lieu" performance hall. In the first, a naked, drunken Dulac entered the stage in a salacious mise-en-scène that drew laughs from the audience until he started cutting himself with a box cutter against the soundtrack of the video circulated online by the killer Luka Magnotta; in the second, which took place during the 2012 student strike, he masturbated in front of bed sheets displaying the name of then province of Quebec prime minister Jean Charest along with a sign reading "Stone me."

Subsequent to these works or performances, several classmates and audience members contacted the school director to either complain about Dulac's cynical attitude or express concerns about the danger he posed to himself. Moreover, the director had summoned the student in winter 2012 to voice the same concerns. Thanks to the testimony of various witnesses, the court also learned that the incriminating project was a cut-and-paste of another project developed for a different course titled "The artist and their career." This course, which aimed to prepare students to build an application file, invited them to "plan a project (fictitious or not), for a targeted area of distribution" and build a portfolio containing, among other things, "a text description of the project" (*TD1*, 235–236). The project could be fictitious or real. It was this project, previously submitted in the above course and unchanged, that Dulac used to apply for the graduate exposition.

The trial was held on July 19, 2013, and Dulac was found guilty of death threats. On the recommendation of his attorney, Véronique Robert, he appealed to the Supreme Court, which overturned the verdict of the lower court on March 14, 2014. He appealed again, claiming the Supreme Court had made a procedural error by failing to correctly consider the analysis of the intentional element of the offence, the *mens rea*. On October 1, 2015, the Court of Appeal accepted the appeal, deeming it legitimate, and overturned the July 2013 verdict. Dulac was acquitted.

The media coverage of the Dulac Affair was so thorough – more than fifty articles were published between 2013 and 2015 in over ten different media –, that few researchers focused attention on it. Four articles briefly analyzed the case: Jonathan Lamy, poet and translator, scrutinized media coverage during the three

---

[4] Duck Hunt is a shooter video game released by Nintendo in 1984 that recreates duck hunting. The player's objective is to kill as many ducks as possible to accumulate points.

years of the trial (2014). In her short article "Avorter. L'oeuvre ou le process" [Abort. The work or the trial], Karine Turcot, a visual artist, examined the different problems posed by the court case ("judging a proposal", "judging beyond the realm", "performance and society") (Turcot 2014, 50). The affair was also the subject of an article by Pierre Rainville titled "Paroles de déraison et paroles de dérision. Les excès de langage à l'épreuve du droit criminal canadien" [Words of unreason and words of derision. Extreme language testing Canadian criminal law], published in *Revue juridique Thémis* a few months before the Court of Appeal handed down its decision in 2015. The article highlighted several inconsistencies in the affair that rendered "the conviction [. . .] wrongful." (Rainville 2015, 50) The law professor emphasized, notably, that the verdict considered neither the context nor the "acknowledged unreality" (Rainville 2015, 83) of the plan to hang children from the ceiling in potato sacks, the absurdity of which should have reasonably removed all doubt about the existence of a threat. Finally, in a short study of the Dulac affair titled "Le parergon mis en procès" [The parergon put on trial], René Lemieux and Simon Labrecque discussed the imbroglio created by the offending work and the way it was tried before the Court of Quebec (with an intermingling of value judgments and quiproquos, according to the authors). (Lemieux, Labrecque, 2014)

For our part, we chose to analyze this case from the perspective of the sociology of art and because it offers a perfect example of conflicting norms. This trial, like many others, can be understood as an effect of the empowerment of art, the empowerment of its values and of its definitions of the beautiful and the good, as opposed to that of common sense on one hand and the rule of law on the other. It also represents a conflict between divergent critique systems, which is at the very heart of the art world. Indeed, the affair began in an art school, where it triggered a conflict not between artists and non-artists, but among the artists themselves. Two levels of conflict must therefore be clarified: one between artistic and non-artistic norms, and the other between artistic norms themselves. To this end, we relied mainly on the work of the art sociologist Nathalie Heinich, notably *Le triple jeu de l'art contemporain* [The triple game of contemporary art] (1998) and *Le paradigme de l'art contemporain* [The paradigm of contemporary art] (2014). Three points will be successively underlined. First, we propose a rereading of the series of reactions to Dulac's projects in light of what Heinich identifies as typical reactions to the conflicts surrounding contemporary art works. Next, we propose two hypotheses to explain why this otherwise banal conflict within the artistic domain changed from an artistic issue to a legal issue: the absence of an explanatory discourse to accompany a creation and the contradictory institutional framework represented by a school of art.

## 2 Conflict of norms: Gallery of typical reactions according to Heinich

In a series of studies about the rejection of contemporary art, Heinich describes what she calls the "triple game of contemporary art", which consists of three main positions: 1) transgression by the contemporary artist, who aims to push moral or aesthetic boundaries, 2) rejection by a public, in the broad sense, which rejects these transgressions on the grounds that they do not constitute art and /or are not morally acceptable, and 3) integration by a public that is sympathetic to contemporary art and always willing to justify these transgressions insofar as the role of art is, precisely, to transgress. (Heinich 1998) According to the sociologist, this triple game continuously widens the gap between the expectations of supporters and opponents: the more the artists provoke, the more their supporters integrate artistic provocations and the more their opponents reject these same provocations. Heinich argues that this is the origin of the long and inexhaustible quarrel of contemporary art that has elicited strong reactions since the 1990s.

Furthermore, this quarrel is fuelled by the coexistence of contradictory definitions of art in the social space and within the art world itself. Heinich distinguishes three paradigms of art that condition practices as well as expectations: the classic, the modern and the contemporary. These three paradigms have appeared successively throughout the history of art, but are today co-present and active, although diversely legitimizing. Much like the mechanics of the triple game of contemporary art (provocation, rejection, integration), the coexistence of these largely incompatible paradigms represents a major source of conflict. We may question some of the generalizations Heinich makes when describing these two key phenomena, but it is likely that both played a decisive role in the Dulac affair.

The transcriptions from the first trial are particularly enlightening in this regard insofar as their 500 pages reveal ten opposing statements. Added to the statement by the accused are those of nine witnesses called on to testify, two prosecutors and the judge. We focus on the witness statements only because, whether for or against, they are particularly recognizable examples of the two typical attitudes of rejection or integration identified by Heinich: attitudes that are, in her view, genuine constants regarding the reception of art and the only ones that support the sociological analysis: "there is no substantiality of tastes, as regards either the object [art] or their subjects [people]: there are only positions, admiring or reactive, positive or negative, integrative or oppositional, regarding fluctuating objects, in varying contexts, on the part of subjects of varying status."[5] (Heinich 1998, 61) The sociologist aims in her work to elucidate "the processes of investment, argumentation and activation of values"[6] (Heinich 1998, 61) in the contemporary regime. To this end, she studied a body of commentaries on contemporary artworks and classified them into four typical groups: rejection by lay persons, rejection by scholars, integration by amateurs and integration by professionals. Note that, based on her observations, the integrative position is characteristic of amateurs and professionals. In other words, in order to integrate contemporary art, one must already be "into it" (*en être*) or, to put it differently, one must already like contemporary art in order to appreciate it. Lay persons, on the other hand, typically take an oppositional position. Heinich's observations can be easily applied to the Dulac affair: matters proceeded as if those involved took it for granted that the oppositional position was typical of lay persons and were careful to avoid the trap of being confused with them.

We'll begin with the integrative reactions, which were also the first reactions to Dulac's text. His project, in fact, was first introduced within a context of complicity during the course "The artist and their career", given by a lecturer.[7] Upon learning of the project, the lecturer found it "funny" and "out-and-out crazy" (*TD1*, 254), exactly as Dulac had expected: ". . . usually, like, she's there and sometimes I mess around with her, I told myself it would be . . . she'd find it funny."[8] (*TD2*, 17) The lecturer was not the only one to have this reaction. Two other students from the same promotion, both called on to testify, responded the same way. Questioned during the hearings, they said they didn't understand what the fuss was about,[9] recounted the audience's fits of laughter during Dulac's performances and insisted it was necessary to know Dulac in order to understand what he did. Added to these

---

5 "il n'existe pas de substantialité des goûts, ni quant à leur objet [l'art], ni quant à leurs sujets [les gens] : il n'existe que des postures, admiratives ou réactives, positives ou négatives, intégratrice ou oppositionnelles, à l'égard d'objet fluctuants, dans des contextes variables, de la part de sujet au statut divers".
6 "les procédures d'investissement, d'argumentation et d'activation des valeurs".
7 The lecturer demonstrated a visible openness to contemporary art: she described herself, moreover, as "a consultant for the ministère de la Culture, in the capacity of specialist in the visual arts." (*TD1*, 229).
8 He specified: "the prof did . . . authorized me to do something with a victim theme, so I thought: "as long as I'm going to do this, I'll exaggerate, really push the boundaries so the meaning is genuinely absurd." (*TD2*, 15) ["le prof faisait . . . m'a autorisé à faire quelque chose de victime, je me suis dit : Tant qu'à le faire, je vais exagérer, pousser ça vraiment à fond pour que le sens soit vraiment absurde"].
9 "I didn't understand what the problem was [. . .]. It didn't surprise me. [. . .] I didn't see anything disturbing about it." (*TD1*, 304–305) ["je comprenais pas il était où le problème [. . .]. Ça m'a pas étonné. [. . .] J'étais pas alarmée par tout ça"].

normalizing reactions was that of the school's visual arts professor.[10] His argument in court was typical of the integrative attitude highlighted by Heinich, which supposes that the purpose of art is not to produce beauty but to encourage reflection:

> David always works with [. . .] taboos, with boundaries in his practice, and this was already the case for [.] his drawing projects, where we saw that, first, there's an object that's very seductive, some very good drawings, and, at the same time something a bit repellent, that creates a kind of malaise, but that's precisely part of his approach. (*TD1*, 266. Our underscores)

> The purpose of David's projects is to work on or issue a critical commentary or to do something meant to be a kind of provocation, just in relation to things acceptably . . . that are acceptable or not in society, but to take those things in society, move them to the art milieu, really specifically to the art milieu, then to work on the meeting between the . . . the norms in the art milieu and the norms in society, to create a kind of . . . of . . . of effect, if you will, of malaise in the viewers, and for him, this is the artistic experience. (*TD1*, 279–280. Our underscores)

Dulac's artistic research would thus involve moving beyond the paradigm of classical art (good drawings, well executed and producing an agreeable effect) to the paradigm of contemporary art, where quality of execution and reception are no longer relevant criteria. In the contemporary regime, the artist is expected to commit a double transgression: moral (Dulac always works with taboos, his works create a "malaise") and aesthetic (something in his drawings is "a bit repellent"). According to the professor's same criteria, a work of art must no longer be evaluated based on the finished product, but on the idea, the concept, the "approach" that led to it, and this approach is characterized, notably, by a blurring of the boundary between art and non-art (Dulac moves *the things of society into art*) and between art and life (art is an *experience*).

This professor's arguments were all the more interesting in that he was visibly working to undo any assumptions the court might have regarding contemporary art. These preconceived notions, which have been widely circulated in the social space since the 1990s thanks to the quarrel of contemporary art, depict the contemporary artist as a worthless provocateur and impostor with little to no talent ("a five-year-old child could have done that"). The professor, therefore, was careful to immediately disqualify the argument depicting contemporary art as transgression for the sake of transgression or as purely negative: Not only did he understate the violence of Dulac's projects (which are "a bit" repellent, which create "a kind of" malaise, "a malaise 'effect'"), but he argued, in particular, that Dulac's aim was not merely "provocation", that his work of negation aimed, ultimately, for something

---

10 He, too, described himself as open to contemporary art, insisting, for example, that his courses on sculpture also dealt with "installations." (*TD1*, 281).

positive, constructive, a "meeting", an "experience." His insistence on using the term "work" and its corollaries when describing Dulac's production returned it to the realm of artistic production that is at once "serious" (not an imposture) and "specialized" (understood by only a special few). This professor portrayed Dulac as a quasi-sociologist, even, and himself adopted a sociologizing perspective in his analysis: it was in terms of *norms* and *milieux* that he defined the dynamic involved in the production and consumption of art.

Another professor presented different authoritative arguments to place Dulac's approach within the realm of art, doing so from the perspective of art history rather than sociology. This was the school director, who evoked the Viennese Actionists, indicating that they were discussed during a "performance class" Dulac had taken:

> . . . the student, quite obviously – this is a bit of artistic information – was strongly influenced by the presentation, in his course, of the work of the Viennese activists [sic]. / The Viennese activists, this is a period of art history when – well I won't give you a lecture on art history – when corporal violence was used to a very large extent; it was a very official movement in the history of art. [. . .] They could wallow in blood, they could do things that were very spectacular and . . . and in bad taste [. . .] that went far beyond the limits of good taste, in the relationship to the body, in nudity, in the relationship to . . . to death up to a certain point." (*TD1*, 170–172).

Both professors were simple witnesses who willingly assumed the role of expert witnesses ("this is a bit of artistic information"), establishing, despite denials, a professorial scenography that created a distinction between those in the know (themselves) and those not in the know (the court) or, more implicitly, "insiders" and "outsiders." Now, they were not alone to further underscore this divide between initiates and non-initiates. The divide, in fact, guided all statements to a very large extent: not only those by witnesses for the defence and the prosecution, but those by the magistrates as well.

The technician, the only witness who was not an artist, insisted he knew nothing about art and was unable to appreciate the artistic value of student projects. Thus, while stating that he found some of Dulac's projects inappropriate, he relativized the relevance of his lay person's opinion and asked that specialists be consulted instead. (*TD1*, 39) The defence attorney acknowledged this same separation between specialists and non-specialists and respected in his own fashion the particular expertise of the specialists from whom he distinguishes himself. He could have emphasized common sense over the rules of art (which the judge would indeed do), but chose instead to play one expert off against another, in this instance, the school director against a professor.[11]

This invisible but active divide between lay persons and specialists placed the artists testifying against Dulac in an uncomfortable position: thus, to avoid being themselves taken for lay persons with no understanding of art, they often reaffirmed their knowledge of art and their experience of its rules. This was particularly evident in the discourse of the lecturer who testified against Dulac and to whom he had given a photograph of his fermented sperm during the "materials" course. While describing Dulac for the prosecution, she characterized herself as hard to "shock" in terms of the projects submitted to her (*TD1*, 57[12]); she was used to the antics of students who, "every year, [. . .] do the same thing" [i.e., plan end-of-session projects that are sexually explicit] (*TD1*, 72). In this regard, witnesses for both the defence and the prosecution stated the same rule: as an artist, in the words of one witness, "we want to keep an open mind" (*TD1*, 277[13]), "we want [. . .] to accept." (*TD1*, 285) Thus, all were aware of Heinich's injunction regarding the integration of transgression by professionals.

It's important to note, however, that some professionals embrace the integrative position unreservedly while others accept it only grudgingly, and that what distinguishes the two is, obviously, whether or not they adhere to the paradigm of contemporary art. Witnesses for the defence included two professors, a lecturer and two students who regularly reaffirmed their interest in installations, performances or videos and willingly represented themselves as mediators of specialized practices. Witnesses for the opposing side included a professor, a lecturer, a student and a technician, who stated they had little knowledge of the history of contemporary art, or demonstrated a preference for modern painting, or frankly admitted they knew next to nothing about art. The explicit argument of the witnesses for the defense mentioned earlier was: *when you know Dulac you don't find his projects disturbing*. Superimposed on this now was another implicit,

---

11 The prosecutor asked this professor if, in his opinion, the director was an artist, and if he knew the codes of the art world ("he understands, the codes, there, when somebody exposes or says something, that he, he understands that we are in an environment . . . a school of art, an artistic environment, he understands that, Mr R." ["il comprend, les codes, là, quand quelqu'un expose ou dit quelque chose, que lui, il comprend qu'on est dans un milieu . . . une école d'art, un milieu artistique, il comprend ça, monsieur R. ?"] (*TD1*, 300) To which the professor replied in the affirmative.

12 "I was quick to react, and I was really very upset, not necessarily by the subject in itself, but to have been . . . well, he didn't . . . he didn't follow my . . . my instructions." Or else "My . . . my students are particularly creative, very open, so, they can come up with proposals that push the boundaries. Which is kind of a good thing, in some way." (*TD1*, 86).

13 The art milieu "is [a milieu] known to be progressive and open." (*TD1*, 277).

but much more active, argument: *when you know about contemporary art, you don't find Dulac's work disturbing.*

It's therefore tempting to read the "misunderstanding" on which the trial is based as a framing issue, as Heinich formulated subsequent to Goffman. The same project, presented twice in two separate courses, received two contrasting receptions: people were amused the first time and found it threatening the second. It all depended on the range of expectation of the interpretive community receiving it. To be sure, the project was presented as "fictitious" the first time and as supposedly genuine the second. But it's also true that the artist and his public shared the same codes, those of the paradigm of contemporary art, whereas, in the second context, the interpretive framework was not defined. And the change of context was, obviously, far more serious when the debate moved from the classroom to the courtroom. Karine Turcot summed up the situation as follows:

> . . . how do you explain to people conditioned to view [. . .] these behaviours as reprehensible "dysfunctions" requiring treatment, even immediate intervention, depending on the seriousness of the action, that strangulation, self-cutting, masturbation and nudity in public, to name only these few, can be considered acceptable, normal, almost banal within the context of a performance? (Turcot 2014, 50)[14]

Nevertheless, an analysis of the different viewpoints during the trial made it possible to temper some of Heinich's observations. The correlation between the presence or absence of insidership in the art world and the type of arguments solicited for or against an artwork appears different in the present affair. According to the sociologist:

> . . . art amateurs, even simple citizens having neither competence nor a particular interest in art, are thus often led to disqualify an artistic proposal, not because of its poor quality or because they don't find it pleasant to look at, but because "it's not art." In doing so, they never or rarely feel that their judgment is illegitimate or inferior. (Heinich 1998, 107)[15]

---

14 "comment expliquer à des personnes dont la formation est de considérer [. . .] ces comportements comme "dysfonctions" répréhensibles nécessitant un traitement, voire une intervention immédiate selon la gravité de l'action, que la strangulation, l'automutilation, la masturbation et la nudité en public, pour ne nommer que ceux-là, sont des comportements pour ainsi dire convenus, normaux, presque banals en performance".
15 ". . . although they are perceived as incompetent, ignorant, even contemptible, by those initiated to contemporary art, they themselves feel, as most of them do, fully authorized to demonstrate their indignation on behalf of the values they defend. So it is no longer unequal stages in the mastery of artistic excellence that are in conflict, but heterogeneous conceptions of what art should be. Thus, the illegitimacy of some is the legitimacy of others, and vice versa, just as positions of domination are largely relative to the value universes in which the actors evolve". (Heinich 1999, 107).

Scholarly rejections, on the other hand, were based not on a criterion of exclusion (such-and-such a work is not art), but on a more tempered approach regarding "the place to assign the work on the scale of relevant values." (Heinich 1998, 227)[16] Now, such profiles were not consistent with those of the lay persons and experts who testified for the prosecution during the trial. This was not insignificant, even if our sample is small. Among those who rejected Dulac's artistic proposals, the single lay person (the technician) outright admitted his incompetence in matters of art, clearly demonstrating "the sense of an illegitimacy, an inferiority of judgment" (Heinich 1998, 227),[17] and two artists – a professor and a student – used the exclusive argument by refusing to accord Dulac's proposals the status of an artwork.

Indeed, there was initially a professor at the school who contested the relevance of one of Dulac's projects on behalf of feminism. This artist then raised an objection having to do with ordinary values, exterior to the rules of art, thus the type of objection that Heinich gladly associates with noninitiates (Heinich 1998, 209), who would be likelier to accuse an artwork of being sexist, as in this case, or perhaps racist or blasphemous. Regarding this specific point, however, the thought occurs that the sociologist's sample was already somewhat dated in 1998 (she referred to a longstanding curriculum) and is still more dated at present: scholarly rejections based on heteronomous criteria have become increasingly common. Indeed, the student who rejected Dulac's proposal on the grounds that it wasn't art was the same one who set the whole affair in motion. She was the very person responsible for collecting the projects for the graduate exposition who, upon learning of Dulac's project, decided "that it was not a project."[18]

When the student found that Dulac's project for the graduate exposition "was not a project", she apparently meant it was not an *artistic* project. Perhaps she also meant it wasn't a "project" insofar as it failed to meet the expectations of the exercise or, more generally, those of a university exercise (she insisted on the fact the work was shoddy). Regardless of the qualification she denied it (this was not an *artistic* or a *university* project, that is, it was not at the artistic or university *level*), the student created a void that could be filled with another qualification: this *non-artistic, non-university* project then became *criminal*.

During the appeal hearing, instead of centering the debate on the *qualification* of the project (artistic, university-level), the defence lawyer focused on the

---

16 "place à accorder à l'œuvre sur l'échelle des valeurs pertinentes".
17 "le sentiment d'une illégitimité, d'une infériorité de jugement".
18 Student A., in her sworn statement during the July 2013 trial, in *The Queen v. David Dulac* [transcription of the trial], Court of Quebec, July 11, 2013, 123. "que c'était pas un projet" See also: ". . . in my opinion, it it's not feasible within the context of the exposition, besides that [. . .] it borders on disturbing." (*TD1*, 134).

*quality* itself, arguing that "it's not a project" in the sense that "it's not a *feasible* project." The project's sheer absurdity, its impracticality, as Pierre Rainville emphasized in an article published some time previously, were indications that Dulac had no intention of taking action, that there was therefore no *mens rea*. It was this interpretation, moreover, that won over the court and led to Dulac's acquittal. Implicitly, however, the perimeter of art was redefined somewhat deviously. Since the project being debated was a purely virtual one, this project-that-was-not-a-project must, according to the defence, have another motivation, which could only be artistic. In an odd reversal, then, which is admittedly fairly common when art shifts from the offensive to the defensive, this project meant to reconcile art and non-art, art and life, was therefore returned to its autonomy.

## 3 On the absence of an explanatory discourse

The project Dulac submitted for the graduate exposition was clearly a draft proposal: "I can't describe the project at *the moment,"* he specified at the start. He would later say in court that he prepared his contribution in a hurry because of his eagerness to take part in the exposition although he had not yet come up with an idea.[19] The organizers' instructions were flexible in that students could hand in an unfinished work: the important thing was to demonstrate their interest in participating. The project design Dulac submitted was consistent with the narrative describing the incriminating text as a work cobbled together at the last minute: the title, "Description of the project for the graduat (sic) Expo by David Dulac", was not exactly a title; it was presented on a handwritten sheet of paper, 8½" by 11", something standard, nothing very fancy" (*TD1*, p 122), recalled the student who received the project, and it was full of spelling mistakes (fourteen mistakes in a 206-word text). As René Lemieux and Simon Labrecque point out, this draft form made Dulac's project potentially ambiguous:

> The word project is problematic here. It can signify two things in this case: either Dulac's "text" represents a future project (he will have the intention, and the text will be proof of premeditation, of kidnapping children and beating them [. . .]); or the text is the project itself, absent the intention of representing a future project. (Lemieux, Labrecque 2014, 20)[20]

---

[19] "I'd gone to see [Student A.], to ask her if I could do what I wanted . . . I was thinking I'd be in the exposition, but I hadn't found anything yet, I hadn't started anything." (*TD2*, p. 19).
[20] "Le mot "projet" est ici problématique. Il peut signifier deux choses en l'espèce : soit le "texte" de Dulac est la représentation d'un projet futur (il aura l'intention, et le texte fera foi de prémédi-

The draft, still in development, was not autonomous; to be comprehensible, additional information was needed from the artist. This necessity was obvious in the Dulac affair.

What made Dulac's work enigmatic, however, was more than just the fact it was unfinished. As Dulac underscored in his final alternative, considerable room was left for interpretation to the point where it was possible to understand both one thing and its opposite: "The aim of the work is to demonstrate how sweet little innocent children will age across the contemporary world into the spineless adults of the future. I of course will represent humanity or its legacy; that depends on point of view." The student, moreover, cultivates a mysterious posture: all his projects provoke disbelief and engage diametrically opposed interpretations (cynicism[21] or denunciation of firearms, complacency or criticism of violence?). Again, this is nothing new according to Heinich: the contemporary artwork leaves its viewer unsure as to its meaning; this is what makes contemporary art an art of initiates or rather, an art that supposes mediation, or an explanation: "No contemporary artwork, irrespective of genre, is presented in the art world without being accompanied by a discourse, whatever forms these 'verbal operators' may take." (Heinich 2014, 175)[22] Accordingly, she refers to an article by Yves Michaud: ". . . additionally, these new forms of art often require the presence of instructions that must be read. We then make the considerable leap from pure contemplation to explanation." (Michaud, Roux 2006)[23] Thus, just as we expect contemporary art to be subversive, we expect to receive explanations or, at the very least, the keys to understand, to decode the work. These explanations may be provided by the artist who created the work, but they may also come from a critic or an exhibition organizer who interprets "what the artist meant" and in so doing, legitimizes the work as relevant and valid.

Now, in this particular case, the explanatory discourse Dulac proposed was inadequate. When inherent in the project, it seemed more like the parody of an explanatory discourse. The first two sentences describe the project, the third indicates its meaning ("The meaning of the work will be [. . .]"). This explanation, incorporated into the text, establishes the cynical character of the project and

---

tation, de kidnapper des enfants et de les battre [. . .]) ; soit le texte est le projet lui-même, sans l'intention de représenter un projet futur."

21 *Cynicism* is used here in its modern sense. It indicates the cynicism of the narcissist who flouts morality without the slightest remorse, with a disarming indifference. *Cf*: Fustin 2018.

22 "Aucune œuvre d'art contemporain, quel qu'en soit le genre, ne se présente dans le monde de l'art sans être accompagnée d'un discours, quelles soient les formes de ces 'opérateurs verbaux'".

23 "ces nouvelles formes d'art requièrent aussi souvent la présence d'un mode d'emploi qu'il faut lire. On passe donc, saut considérable, de la pure contemplation à l'explication."

apparently plays with the conventions of contemporary art. By pluralizing the levels of meaning in this way, Dulac highlighted the conflict between social norms, artistic norms and, further still, the norms governing the art world itself. In the paradigm of contemporary art, Heinich reminds us, the explanatory discourse assumes the form of an interpretation both in its hermeneutic sense by imputing meaning and in its pragmatic sense by executing the work, as is said of a musical production (Heinich 2014, 190); the public relies on it to access the work. That Dulac accompanied his project with an explanatory discourse confirmed the nonsense, the absurdity of the work and created an illusion of interpretation. He moved between different levels of meaning and took pleasure in challenging them, confusing them.

When Dulac was asked to explain his text in court, he gave a more comprehensive explanatory discourse, indicating that his ideas came from the ambient discourse; that, because the professor had given complete freedom for the project design, he had wanted to take advantage of this freedom and "exaggerate the nonsense", push his project "to the limit to render it genuinely absurd" (*TD2*, 15). As a result, he had drawn inspiration from common parental warnings to children about not getting into a car with strangers. Regarding the second part of the project evoking children in potato sacks hanging from the ceiling, he explained the idea had come from an Austin Powers movie in which "Doctor Taylor talked about his youth and he . . . something like that happened to him. I, like, added the sacks (*TD2*, 13)." When Dulac tried to reveal and describe his references, explain his approach, he was visibly uncomfortable. It was obvious he found the exercise painful, since a difficulty of this kind is highly prejudicial in a context where the explanatory discourse represents the completion of the work and determines the shift towards art or non-art.

Although he never produced an explanatory discourse in due form, Dulac seemed at least intuitively aware of the triple game of contemporary art, notably the injunction artists must transgress. When the art school director summoned him in winter 2012 to explain that certain department members found his project disturbing and he was playing dangerously with "alarm signals", Dulac answered: "I thought that was the idea." (*TD1*, 192) Indeed, as Heinich recalls, transgression in art "is not to be confused with the absence of norms; nothing is more norm-governed or more contrived than the work of an artist who seeks to overstep the boundaries without, however, being excluded, to modify the rules of the game without, however, being declared out of line." (Heinich 2014, 56)[24] It was obvious

---

[24] "ne se confond pas avec l'absence de normes ; rien n'est plus normé, plus contraint que le travail de l'artiste qui cherche à franchir les limites sans être pour autant exclu, à modifier les règles du jeu sans pour autant être déclaré hors-jeu."

that Dulac had not managed to master these norms well enough "to not be declared out of line."

Whether or not he had mastered the rules of the game and delivered a convincing explanatory discourse, it must be noted that Dulac did not enjoy the notoriety required for effective self-authentication. What he needed was the legitimizing discourse of a recognized third party (a peer or a professional). Now, this legitimizing discourse had not been delivered before the situation deteriorated, and it could be heard only in court. It was articulated during the trial in the form of testimony by one of the school's professionals who ran the contemporary art program.[25] This professor then returned, as it were, to Dulac's refusal of an explanation and the very essence, in fact, of his artistic gesture:

> It's important just the same, [. . .] to understand these nuances, that are there, in different ways of remaining silent, it's that, when artists want their silence to act upon their work, the fact they don't speak becomes a gesture, [. . .] if we don't expect somebody to speak when something is puzzling or when we don't know how to understand it, and the person says nothing, we tend to be surprised by the silence. [. . .] David's silence is part of his work, in the same way that someone else's discourse may be part of their work. (*TD1*, 276–278)

With this position, the visual arts professor not only explained to the court the different postures that artists can adopt, he delivered the missing explanatory discourse as well. He clarified the effect of silence on the work, this "effect of dramatic tension" that characterized the incriminating work and, in his capacity as a specialist, highlighted the artistic approach that underlay all the student's productions.

He began by positioning himself straight away as a specialist in Dulac's work, saying: "while observing his practice throughout the [. . .] bachelor's degree, and also through discussions with him, and also through his presence in group discussions, group critiques, I really came to understand [. . .] his artistic approach, then his artistic process and so I can talk about it." (*TD1*, 262–263) Based on this authority,[26] he then made it a point to declare that "David's silence is part of his work" – thereby confirming the student's coherent and creative artistic approach. He explained that the reason for Dulac's withdrawal and lack of an explanation was to allow viewers the freedom to interpret his work, to determine its meaning for themselves. According to the specialist, the artist wanted to place the responsibility for an explanatory discourse in the hands of the public.

---

25 Courses on theory, sculpture, drawing. (*TD1*, 261).
26 Nathalie Heinich maintains, moreover, that the discourse of a specialist is the consecration of the artist: "Success in contemporary art involves the discourse of a specialist." ["La réussite en art contemporain passe obligatoirement par le discours d'un spécialiste".] (Heinich 2014, 180).

In short, the Dulac affair epitomizes the cruel rule that makes the triple game of contemporary art a game of double or nothing. In debt of recognition, the illegitimate artist who engages in contemporary art greatly risks increasing their illegitimacy. This is precisely what occurred in this instance. Dulac failed to produce a convincing explanatory discourse defending his work; as well, he failed to build a legal defence that managed to convince laypersons that the rules of contemporary art are especially dangerous for artists who engage in it. A contextual element also demonstrates why the rules of the artistic game, already unforgiving, were difficult for Dulac to follow: they were at variance with a normative system that was just as prevalent – that of the university where the affair took place.

## 4 On the conflict between artistic and university norms

Derived from the Latin word *universitas* designating a "community, corporation", the university today is an institution of higher learning "composed of a set of didactic organizations."[27] Each of these organizations has a certain personality and develops its own normative framework. In this case, the Dulac affair developed within the Laval University school of art. Now, a certain discrepancy is observed between the University's general statement to the effect that its main values are responsibility and respect[28] and the presentation of the art school itself, which is advertised as "a dynamic environment among the most challenging, and one in constant evolution", one whose "objective is the development of individuality in each student, based on their world view and particular experience, and the expression of these across a privileged artistic domain."[29] Although they are not entirely contradictory, the first vision is more conservative and focused on living together than the second, which emphasizes movement and original self-development.

---

27 "constitué par un ensemble d'organisations didactiques". "Université", *Grand Robert de la langue française*, [on line].
28 (s. a.), "Mission, vision, valeurs", *Université Laval* [on line], URL: https://www.ulaval.ca/notre-universite/mission-vision-valeurs.
29 "un environnement dynamique des plus stimulants, et en évolution constante" and "objectif est le développement d'une singularité chez chaque étudiant, basée sur une vision du monde et un vécu particulier, et leur expression à travers un domaine artistique privilégié". (s. a.), "Présentation", *École d'art. Université Laval* [on line], URL: https://www.art.ulaval.ca/a-propos/presentation.

As was clearly articulated by the art school director invited to testify in court, a school of art remains a place of supervision that must remain safe.[30] Art schools and departments thus share a particular feature: they attempt to establish a framework and norms in a domain where, very often, these same norms are constantly tested and resisted. On one hand, the school teaches that playing with the rules of common sense is a current practice in art, leading Dulac to do this very thing without asking questions. On the other hand, it expected him to obey the rules of common sense, causing him to be reproached for applying theory learned in the classroom. The art school director himself oscillated between the position of artist, who understands the transgressive injunction and that of director, who guarantees safety in the school. He recounted that he tried at times to "bring [Dulac] onto the terrain of art" in order to "create a dialogue" and at other times, conversely, reminded him to respect the institutional norms of living together: "Well, look, it's because, in a school, you're not allowed to create an environment that's . . . that's disturbing for your colleagues. You just can't do that." (*TD1*, 191–192) The director's hierarchical authority was obvious in this tension between divergent norms: when asked to explain the significance of the expression "alarm signals", he offered the following quasi-semiological definition: "There are gestures, words, language systems whose meaning differs from one area to another." (*TD1*, 195) In other words, one has to know how to handle the codes, even when they are intrinsically contradictory.

It must be noted here that several persons in the affair were acting in a dual capacity: they were both teachers and artists, which blurred the boundary between the "terrain" of art and that of the university. This was even more so given that most of the teachers in the art school were employed there precisely because they had a parallel artistic practice. The school's evaluation criteria were similarly based on artistic relevance in the world of contemporary art. The whole process of selecting works for the final exposition corresponded perfectly to the norms in effect in the art milieu. Finally, as was learned from witnesses during the hearings, certain venues for the artistic performances of students in the school were both located outside the school and indirectly linked to it. Such was the case for the "Lieu", the site of Dulac's presentation when he performed naked and self-cut with a box cutter. The founder and director of the multidisciplinary art centre, established in 1982 in the St-Roch area of Quebec City, was also a lecturer in the art school. Additionally, this venue exemplified the blurring between the art school and the art milieu, as one student testified during the hearings:

---

[30] "[R]: my objective is for the school to be a proper place to work in. [Q]: a safe place. [R]: well, yes." (*TD1*, 220).

> Because our . . . our performance course professor is Mr. M, who is also the director of the Lieu, the performing art centre, so, for him, it made sense that we . . . for our exam to be given right in a context of performative art, right there, and with . . . with an initiated public, and finally, with a public initiated to that, so, for him, it was interesting to really place us in that context for our . . . our evaluation, instead of simply in an academic context, because there would also be an initiated public. (*TD1*, 318)

Who, then, should Dulac obey? And what norms should he obey: those of the university institution or the art world? The question becomes even harder when these contradictory norms are embodied by the same persons and the same institutions. Within the university itself, the creative vocation of the art school, the composition of its personnel and the values disseminated in it complicate the rules of the university game and its specific injunctions. As Heinich shows in *The Triple Game of Contemporary Art,* schools of art are "one of the touchstones of institutional action by which artistic innovation manages to integrate into transmissible heritage, at the cost of a more or less drastic redefinition of the acceptable and, from there, of the teachable." (Heinich 1998, 281) They are places of transmission where "the issue is not so much to provide the historical culture that makes it possible to give meaning to artists' proposals, but to 'sensitize' young people and stimulate their creativity by connecting works to their lived experience." (Heinich 1998, 281) This connection cannot be made without the conflicts of norms that often play out on a personal level: students must learn to maneuver, as Sarah Mecarelli demonstrates, in order to "Construct a field of adequate cultural and theoretical references" and "Know how to talk about their work and formulate a narrative."[31] Mecarelli thereby accounts for the set of formal and informal dynamics between students and professors that characterize the training of artists and reveal the complexity of the power relationships at play.

Thus, one sees in the Dulac affair a manifestation of the permissive paradox identified by Heinich. Institutions, historically called on to establish boundaries between art and non-art, modify their vocation in the contemporary regime. Instead of establishing a framework boundary, they broaden the framework. They defuse, in some way, the transgressive charge of provocative works, extending the boundaries of what is acceptable.

> For as soon as it is confined to spaces that are (increasingly) reserved, contemporary art loses its provocative power, eroded in advance by the routinization of initiates and the quasi-immediate acceptance of institutions by virtue of the "permissive paradox", which makes art intermediaries who are in charge of managing artistic boundaries the active accomplices of artistic transgressions; in this way, institutional permissiveness forces artists to make greater and greater efforts to expand limits that are pushed back indefinitely. (Heinich 2015, 134)

---

31 Mecaralli Sarah, in Heinich 2014, 178.

## 5 Conclusion

Thus, in the Dulac affair, a few lines in a text were endlessly interpreted and reinterpreted. Events proceeded as if the vacuum left by the absence of an explanatory text by the artist called for interminable explanations aimed at establishing the tenor and meaning of this unidentified object: Is it art? Is it a university assignment? Is it serious? Is it threatening? Is it normal? We've seen to what extent the answer to this last question depends on the person expressing it, and sometimes even on the role that person feels obliged to assume. An example is the professor whose responses differed based on his self-assumed role as art historian or department head.

Confronted with such a wide variety of receptions, we understand the interest in the judicial notion of "reasonable person", since the judge must adopt the viewpoint of this "person" to decide whether or not a discourse is threatening. The issue here, according to doctrine, is the average citizen who is "reasonably informed" (Quebec 1993, 785–786),[32] but who, at the same time, has no conflict of interests or particular knowledge of the subject concerned and is therefore at once "objective" and "secular." (Chênevert 2015, 31) In short, the issue is not the institutional permissive ready to integrate all proposals because they are artistic, nor the fierce rejection of contemporary art by those flatly uninterested in explanations of its codes and its history. Indeed, if we don't want this "judicial fiction"[33] to conjure up the outdated notion of the "good father", or even of the judge him or herself, if we are to prevent its transformation into "law's ghost god" (Duhaime, 2012), we must remember that, according to case law, the reasonable person must be not only "neutral and informed", but also "well aware of all the circumstances" and "required to study the question 'in depth'." (Rainville 2015, 113)[34] He or she must take into account, notably, the "manner in which [the words] were uttered and the person for whom they were intended." (Rainville 2015, 113)

As Pierre Rainville stated, while pointing out that "the reasonable person should not be 'introduced to artistic activity'" and then dismissing "the opinion of artists that the intended meaning is not intimidating"[35] (Rainville 2015 82), both

---

[32] "normalement avisé".
[33] *Ibid*, 1.
[34] He cites the Batista judgment, reprised in R. v. McRae, n 107, par. 14–16. "To decide if a reasonable person would have considered the words uttered to be a threat, the court must examine them objectively, while taking into account the circumstances in which they occurred, the manner in which they were uttered and the person for whom they were intended. Evidently, words uttered in jest or in a way that cannot be taken seriously could not lead a reasonable person to conclude they constitute a threat." (R. v. Clemente, [1994] 2 R.C.S. 758, 763).
[35] "initiée à l'activité artistique", and "l'avis d'artistes selon lesquels la signification recherchée n'est pas intimidante".

the judge at the first instance trial and, later on, the Supreme Court blocked the possibility of correctly elucidating the context of the offending words and therefore making an informed decision about the threat. What's more, one wonders if the elucidation of this context was not good enough reason to reconsider the *mens rea* element, since Dulac's intention was further clarified in light of the contradictory injunctions to which he was submitted by setting foot in a school of art, like an electron released into a force field.

## Bibliography

Chênevert, Paul. 2015. *La variabilité du concept de personne raisonnable dans les décisions de la Cour Suprême du Canada*, Thesis in law, Université Laval, Québec.

Duhaime, Lloyd. 2012. *The Reasonable Man – Law's Ghost God, Law Dictionary and Legal Information* [on line], URL: http://www.duhaime.org/LegalResources/TortPersonalInjury/LawArticle-1378/The-Reasonable-Man-Laws-GhostGod.aspx. (13-08-2022)

Fustin, Ludivine. 2018. *Cynisme, parrêsia et scène littéraire*, Poétique 183(1), 23–38.

Heinich, Nathalie. 1998. *Le triple jeu de l'art contemporain*. Paris, Minuit.

Heinich, Nathalie. 2005. *L'art du scandale. Indignation esthétique et sociologie des valeurs*, Politix 71(3), 121–136.

Heinich, Nathalie. 2014. *Paradigme de l'art contemporain*. Paris, Minuit.

Judgment, *The Queen v. David Dulac*, Court of Québec, July 19, 2013.

Labrecque, Simon & René Lemieux. 2014. *Le parergon mis en procès*, Raisons sociales [online], URL: http://raisons-sociales.com/articles/parergon-mis-en-proces-juridique-prend-limposture-au-serieux-remy-couture-david-dulac/ (13-08-2022)

Lamy, Jonathan. 2014. *Chronique d'une condamnation annoncée*, Inter. Art actuel 117, 49–51.

Michaud, Yves & Emmanuel Roux. 2006. *L'art en mutation*, Le Monde [online], (URL: https://www.lemonde.fr/culture/article/2006/05/20/yves-michaud-evoque-l-art-contemporain-du-xxie-siecle_774107_3246.html) (13-08-2022)

*Mission, vision, valeurs*, Université Laval [on line], URL: https://www.ulaval.ca/notre-universite/mission-vision-valeurs (13-08-2022)

*Présentation*, École d'art. Université Laval [on line], URL: https://www.art.ulaval.ca/a-propos/presentation (13-08-2022)

Québec (province), *Civil Code of Quebec: Commentaires du ministre de la Justice, Tome 1*, Québec, Gouvernement du Québec, 1993.

Rainville, Pierre. 2015. *Paroles de déraison et paroles de dérision. Les excès de langage à l'épreuve du droit criminel canadien*, Revue juridique Thémis 49, 35–132.

*The Queen v. David Dulac* [trial transcription], Court of Québec, July 11, 2013.

Turcot, Karine. 2014. *Avorter. L'œuvre ou le procès*, Inter. Art actuel, 118, 50.

"Université", *Grand Robert de la langue française*, [on line]. (13-08-2022)

Stanisław Goźdź-Roszkowski

# Argumentation, rhetoric and legal justification. The case of Poland's Constitutional Tribunal ruling on abortion

**Abstract:** This chapter examines the justification of the controversial abortion ruling given by Poland's Constitutional Tribunal from the perspective of legal argumentation theory and incorporating certain elements of classical legal rhetoric (*ethos*, audience effect, *topoi* and *kairos*). Judicial reasoning is reconstructed using weighing and balancing as a principle method for external justification but other argumentative devices identified as relevant include arguments from authority and the use of emotive, value-laden language. The analysis supports the view that argumentation contained in the opinion is unique since it needs to draw on both interpretive as well as rhetorical methods far more often than normally would be the case. The rhetorical elements help to account for the impact of the political, legal and social contexts in which the ruling is embedded. It emerges that irrespective of the argumentative merits of the justification, it cannot be considered as effective and acceptable because of the court's undermined legitimacy and a failure to convince key audiences.

**Keywords:** abortion, argumentation, justification, persuasion, rhetoric, value-laden language

## 1 Introduction

Justifying decisions regarding morally sensitive issues, such as abortion, euthanasia or medically-assisted procreation, belongs to one of the most difficult tasks faced by judges. In such cases, judges are expected to rule unequivocally and rationalize matters which do not lend themselves easily to rationalizing. Whether international or domestic, adjudicating bodies must find solutions to fundamental conflicts of incommensurable constitutional principles. This inevitably involves

---

**Acknowledgment:** Research reported in this chapter was supported by National Science Centre Poland under award number UMO-2018/31/B/HS2/03093.

---

**Stanisław Goźdź-Roszkowski,** University of Łódź

https://doi.org/10.1515/9783110799651-003

balancing rights, values and goods, which in the context of terminating pregnancy, often boils down to considering the dichotomy between a right to life and a right to self-determination, the bedrock of the human rights order. The open-ended nature of human rights precludes relying solely on standard methods of pure logic and legal syllogism. Judges may feel compelled to incorporate rhetorical methods into their reasoning in order to explain and justify their decision rather than merely interpret legal norms. Presenting a solution based on a purely logical analysis may simply turn out to be insufficient in the face of a multiple audience awaiting not only the outcome of the judicial decision-making but also the motivations behind it. Court cases deciding morally sensitive issues, especially those concerned with expanding or limiting the right to abortion receive intense media attention evoking a range of emotive responses from the audiences.

The position of adjudicating judges may be made even more precarious when the court is facing serious doubts about its legitimacy. Worse still, the setting and the time when a crucial ruling is announced may suggest that the court is deeply entangled in the conflicted political realities as it becomes a plaything in the hands of populist politicians. All these factors had substantial bearing on the Polish Constitutional Tribunal's decision which ruled on 22 October 2020 that a clause allowing the termination of pregnancy in the event of a fatal foetal abnormality was unconstitutional. As a result, abortion law in Poland has become one of the strictest in Europe permitting abortion in only two cases: a threat to a woman's life or health and in the case of a rape or incestuous pregnancy. The ruling provoked harsh criticism and a public outcry as a blatant violation of human rights; all of this occurring at the peak of the coronavirus pandemic in Poland.

The purpose of this chapter is to shed new light on the use of argumentative and rhetorical devices to justify the expansion of one right and the limitation of the other in the context of the controversial abortion ruling. This is done by drawing upon legal argumentation theory, mainly Alexy's Theory of Balancing (2002; 2003) but also identifying common lines of argument (*topoi*) and arguments from authority. In addition, it is argued that the justification of the ruling cannot be accounted for without giving some attention to certain aspects affecting the Tribunal's decision that are more likely to be discussed in research on legal persuasion, rhetoric or political communications (e.g. Berger and Stanchi 2018; Romano and Curry 2020; Frost 2016).

In the following, I first present an overview of legal justification perceived as argumentative activity explaining why rhetorical and dialogical approaches are also relevant to the study of legal justification. The discussion then starts with examining the macrolevel aspects of the justification: the credibility of the adjudicating panel, the audience effect, and the time of announcing the ruling. These aspects also serve as a background to the analysis of the text of the justification in light of the legal argumentation theory. Finally, I reflect on the relevance of my results to broader issues of how interpretative, argumentative and persuasive concerns are closely intertwined in legal justification.

## 2 Legal justification as argumentative activity

While judges enjoy a certain degree of leeway in interpreting and applying the law, they are expected to account for how they use this discretionary space. This usually achieved by advancing argumentation in which judges justify their choice of a specific solution against the backdrop of various rules and forms of legal interpretation, the application and creation of legal rules valid under a given legal system. The perception of law and legal justification as an inherently argumentative action has been acknowledged and documented in numerous publications and led to the emergence of different theories on the justification of judicial decisions (e.g. MacCormick 2005; Alexy's Procedural Theory of Legal Argumentation 1989; 2002; see also a survey of argumentation-oriented theories in Feteris 2017). These led to the perception of law as essentially arguable in character and the view of legal argumentation as no longer defined by "the certainty of demonstrative arguments with undisputed premises and deductive proof: legal questions are answered in terms of acceptable argumentation (Kloosterhuis 2013, 71)." Judges are expected to convince their audience that their decisions are acceptable in terms of the norms and standards provided for the application and interpretation of the law (Feteris 2017; for a discussion of the concept of judicial audience and its relation to acceptability, see Perelman (1969) and Aarnio (1987).

The argumentative nature of legal justification seems to be particularly conspicuous in cases regarding colliding or competing principles and rules. If the principles concern morally sensitive issues such as the right to abortion, the legal outcome is seldom to be discovered "by means of an argument that reconstructs the operation of legal rules" (Hage 2013,126). Rather, the outcome hinges upon those arguments that are selected and used in the legal reasoning. As Hage (2013,126) emphasizes "the legal consequences of a case are constructed by means of the arguments, and not merely reconstructed. The legal consequences of a case would then be what the best (possible) legal argument says they are".

Thus, legal argumentation should be viewed as performing an essentially persuasive function on a different level. Judges who are engaged in legal reasoning draw upon certain methods of interpreting norms and the interpretation that obtains performs an argumentative role because it should be acceptable by a legal audience or audiences that it is directed at. The rhetorical approach to legal argumentation shifts attention from the logical approach with its focus on formal aspects towards the content of arguments and the context-dependent aspects of acceptability (Feteris and Kloosterhuis 2013; see also Toulmin 1958). Arguably, the acceptability in cases involving the open-ended norms regulating human rights is dependent on the effectiveness of the argumentation for the audience to which argumentation is addressed. This calls for a more dialogic and argumentative style of reasoning and justification in which judges make references to values, principles and interests, which then need to be weighed and balanced to convince people of their decisions.

## 3 *Ethos* or the credibility of the adjudicating panel

Under classical legal rhetoric, the effectiveness of legal argumentation hinges upon the speaker's credibility and the strength of his or her character (*ethos*) just as much as on logical integrity (*logos*) and emotional content (*pathos*). The importance of *ethos* in modern trial advocacy handbooks is usually discussed in the context of lawyers' credibility when arguing their cases before lay juries (Frost 2016). But what if the concept is applied somewhat differently to see whether the credibility of a judicial institution has not been undermined in the eyes of legal and non-legal audiences?

Viewed from the perspective of a legal academic community in Poland, the Constitutional Tribunal's ruling should not be even considered as valid because of some formal and procedural defects originating with the institution itself (Gliszczyńska-Grabias and Sadurski, 2021). As some would point out, the ruling was handed down by a panel of judges which included judges who had been improperly appointed and were not qualified to adjudicate. The same charge of improper appointment is raised with regard to the president (chief justice) of the Constitutional Tribunal. In addition, the ruling was published well after its statutory deadline, which breached a constitutional requirement. In Poland, constitutional court rulings must be officially published to take effect. Other legal experts challenged the ruling arguing that some of the judges were involved in drafting and submitting the application to the Constitutional Tribunal thus violating the *nemo iudex in causa sua* principle. The last point was raised in one of the dissenting opinions (Judge Leon Kieres in K1/20, Sec. 2.2).

The erosion of trust in the Constitutional Tribunal as an independent judicial institution should be seen against a wider backdrop of the rule of law crisis in Poland (Sadurski 2019a). Soon after its landslide victory in the general election in

2015, the new ruling party embarked on a process of overhauling the Polish judicial system starting with the Constitutional Tribunal. The outcome of this process is succinctly assessed in the following passage written by a leading authority on constitutional law:

> In Poland, the Tribunal became a defender and protector of the legislative majority. This changed role, combined with general distrust of the Constitutional Tribunal and concerns about legitimacy of its judgments, explains also the extraordinary drop in the number of its judgments. For all practical purposes, the Constitutional Tribunal as a mechanism of constitutional review has ceased to exist: a reliable aide of the government and parliamentary majority has been born (Sadurski 2019b, 63).

There are other audiences beyond the legal community, however, and these also play a role in how the Tribunal has been perceived. According to a public opinion poll published in December 2020, 63% of Poles supported protests against the ruling.[1] In addition, 50% of the respondents expressed a negative opinion about the Tribunal's judicial activity. This marks a clear drop from a public opinion poll conducted in 2015 when the overhaul of the judicial system started. At that time, only 12% of the respondents expressed a negative sentiment towards the court, while 42% viewed it favourably. It appears that the prestige and credibility of the Constitutional Tribunal has rapidly eroded within the space of a few years casting a long shadow over the legitimacy of its rulings.

## 4 Audience effect and the timing

According to the rhetorical approach, attention should be given to the content of arguments and the context-dependent aspects of their acceptability (Feteris and Kloosterhuis 2013). One such context-dependent aspect is the audience. The acceptability of argumentation is conditional upon the effectiveness of specific arguments that are selected with a view to persuading a specific audience or audiences. Such persuasive arguments are designed and deployed to address the concerns and priorities of those that one is trying to persuade (Clark 2013). In the context of majority opinions, drafted by supreme courts or constitutional courts, judges need to address a composite audience (Makau 1984) consisting of the fellow judges ('the bench'), lower court justices, litigants, legal community, legislators, government of-

---

[1] CBOS (Eng. Centre for Public Opinion) 2020, 159/2020, https://www.cbos.pl/SPISKOM.POL/2020/K_159_20.PDF (last viewed 20.08.2022).

ficials, but also news outlets and the public. It is the last two audiences that are most likely to be interested in the outcome of a much-publicised case (see also Romano and Curry 2020, 27–33) for a discussion of what communities can be subsumed within the concept of judicial audience).

Clearly, the composition of a particular judicial audience will depend on a legal culture and a given court but the key issue in legal justification is to identify who is to be convinced by the argument. If the rule of law remained intact in Poland and its constitutional court was not captured and abused by the populists (Kovalčík 2022; Koncewicz 2018), then discussing the audience effect regarding the ruling could take a different direction and focus on efforts towards achieving recognition and maintaining the legitimacy of judicial decisions. Instead, it appears more relevant to view the ruling and its justification as an action undertaken at the ruling party's instigation. According to some commentators (e.g., Wigura and Kuisz 2020), the ruling was a cynical attempt by the government to maintain the support and loyalty of the most conservative Catholic voters, many of whom are residing in rural areas. It may also have been announced to distract the public from the government's failure to tackle the effects of the Covid-19 pandemic. Whatever the real reasons were, in that case, bowing to the political pressure, the Tribunal's opinion was written with a very limited range of audiences in mind. Interestingly, the arguments used to support the ruling did not persuade even some of the judges from the bench and as many as five separate (dissenting and concurring) opinions were released, somehow undermining the certainty of the decision.

While, arguably, not a decisive factor, there is something to be said about the temporal aspect of the Tribunal's ruling. The application for a constitutional review of Art. 4a (Sec. 1, 2) of the *Act on Family Planning* regarding foetal impairment was first submitted in October 2017 by a group of right-wing MPs. At the same time, a draft bill to amend the Act was introduced to the lower chamber of the Parliament. The parliamentary work was carried out for some time but it was eventually discontinued. Similarly, the application to the Constitutional Tribunal was frozen only to be revived in practically the same form soon after the general election was won by the ruling party. The science of legal persuasion recognizes that it often takes time to prepare the grounds and pick the right moment for coming up with a case (Berger & Stanchi 2018). Whether or not the ruling party and its 'puppet' court were thinking in terms of *kairos* and were searching for an opportune moment to announce the ruling, their plan badly backfired. As soon as the ruling was announced, mass protests erupted across the entire country, despite restrictions on the right to assembly in place due to the COVID-19 pandemic. As a result, the ruling came into effect belatedly in January of 2021.

With this background in mind, we can now proceed to examine the text of the majority opinion with a view to identifying and analysing major arguments used to justify the ruling.

# 5 Commonplaces or *topoi* in the Tribunal's reasoning

Commonplaces (also referred to as *topoi*, or *loci communes*) are identified in the classical legal rhetoric as common lines of argument or 'argumentative sites' which need to be identified and described as argumentative premises to ensure the completeness and effectiveness of the ensuing reasoning (Frost 2016). Thus, they are usually used as starting points referring to values commonly accepted in a given legal and societal order and likely to be considered persuasive. Such values are projected as fundamental to the legal and social order and in need of protection. Legal topoi usually draw upon general legal principles, such as fairness or equity (Feteris and Kloosterhuis 2013). Viewed from the rhetorical perspective, they are akin to ethical or emotive words, which are characterised by the close relationship between value judgments and emotions (Macagno and Walton 2016, 31). They can become a powerful instrument of persuasion because they *are* effectively value judgements and are thus capable of affecting attitudes.

It is human dignity that serves as the starting point and commonplace in the majority reasoning. Citing its previous case law, the Tribunal asserts that human dignity as such is not a right. Rather, it is the foundation and the source from which rights are derived. Since dignity is the source of rights, it cannot be in any way restricted by them. It is a point of reference for the constitutional system of values and the entire legal order. The Tribunal then continues to justify the superior position of dignity: "Dignity is inherent and inalienable. As such it does not require normative legitimacy because its source does not obtain from any entitlement bestowed by the state. In addition, dignity is primary and it constitutes an opening of the constitutional system to extra-legal values" (K 1/20, Sec.3.1).

The majority opinion applies a value hierarchy by positioning human dignity at the top of the constitutional axiological system, emphasizing the priority of human dignity within the constitutional axiological system and projecting it as a value that needs to be respected and protected (K 1/20, Sec. 3.1). The Tribunal establishes then a direct link between human dignity and the legal protection of life. Its position is clarified by referring to its previous opinion that "it is not possible to talk about protecting human dignity if there is no sufficient basis for the protection of life".[2] Establishing a causal link between the protection of dignity and the protection of life is possible by modifying the meaning of these terms. This is possible given the open-ended and general nature of *topoi* as well as the generality of the constitutional provisions related to these concepts. The majority opinion distinguishes two aspects of human dignity. First, it argues that dignity can be understood as an inalienable and inherent human value. Second, it could also be associated with the "value of human psychological life and all those values that define an individual's position in society and which are worthy of respect"[3] Using evaluative language, the Tribunal expresses its preference for the first understanding of dignity because, in its view, it has its roots in the existence of a human being and not merely a sense of well-being.[4]

This illustrates the central importance of definitions and how modifying a definition can amount to an act of persuasion. The Tribunal opts for a specific sense of the term *dignity* to equate it with the concept of *right to life*. However, a close reading of Article 30[5] shows that dignity is a source of *all* freedoms and rights and it does not apply solely to *right to life*. In addition, if dignity itself is a source rather than a right, then it cannot be invoked as a basis for the legal determination of a right. Instead, it can only be used as an "interpretive tool for rights analysis" (Gliszczyńska-Grabias & Sadurski 2021, 141). In the context of this case, the choice of such interpretation of dignity carries important implications for other constitutional rights, especially the right to equality before the law and non-discrimination (Article 32), bodily inviolability (Article 41), and privacy (Article 47). There is nothing in the wording of Article 30 to suggest the primacy of right to life. Further, the way the majority opinion interprets dignity does not seem to include the dignity of women. This is surprising even on formal grounds because the issue of the relationship between the rights of the pregnant woman and the rights of the human being in the prenatal phase of its life was raised in the motion for constitutional review and by the *amici curiae*. This omission is actually noticed in one of the dissenting opinions: "The Constitutional Tribunal (. . .)

---

[2] The legal protection of life is enshrined in Art. 38 of the Constitution which reads: 'The Republic of Poland shall ensure the legal protection of the life of every human being'. The English translation of the Polish constitution used in this chapter comes from the official website of the Polish parliament available at https://www.sejm.gov.pl/prawo/konst/angielski/kon1.htm (last visited 16.08.2022).

[3] Constitutional Tribunal's ruling of 5th March 2003, case no. K 7/01. The excerpts from the Constitutional Tribunal's opinions used in this chapter have been translated by the author, unless otherwise indicated.

[4] Constitutional Tribunal's opinion of 5th March 2003, case no. K 7/01.

[5] Article 30 of the Constitution reads: "The inherent and inalienable dignity of the person shall constitute a source of freedoms and rights of persons and citizens."

only considered one perspective—the perspective of protecting life in the prenatal phase. At the same time, it ignored the perspective of women whose dignity, life and health are undoubtedly values under constitutional protection."[6]

The way in which the majority opinion interprets the concept of dignity, as inextricably linked to the existence of a human being, relegates other aspects of dignity to an inferior position, enabling the Tribunal to disregard the dignity of women. The use of emotive words can indeed be also seen in the dissenting opinion, which refers to value-laden lexical items such as *heroic attitude, life and health, sacrifices and hardships*:

> In the name of protecting life in the pre-natal phase—which is not absolute—the Constitutional Tribunal imposed on women the obligation to have a heroic attitude, i.e. the obligation to assume responsibility in all circumstances—regardless of the nature and degree of pathology of the foetal development or the possible consequences for their life and health with the continuation of pregnancy—for sacrifices and hardships far exceeding the usual measure of limitations related to pregnancy, childbirth and raising a child.[7]

The concept of life also required some modification on the part of the Tribunal. In a rare originalist turn, the Tribunal referred to the files of the Constitutional Commission of the National Assembly, which was tasked with drafting the Constitution of 1997, to determine the legislator's intention regarding the meaning of the term *human* (Section 3.3.3). The outcome of this reference is surprising. Rather than use the files to disambiguate the meaning of the term, the Tribunal construed the politically-situated parliamentary debate to assert its authority to resolve the matter (K 1/20 Section 3.3.1): "Given the foregoing, the Tribunal accepts that it was the legislator's intention – in the absence of any express reference to the temporal limits of human life and its protection in the Constitution (. . .) – to leave it up to the Tribunal to determine the meaning of *human* subject to Article 38 of the Constitution."

This curious move has attracted criticism about the Tribunal's undue judicial activism, to the detriment the legislative power (e.g. Bucholc 2022; Gliszczyńska-Grabias & Sadurski 2021,141). In its conclusion, the majority opinion defines human life as a value at each stage of its development and the value derived from the constitutional provisions must be protected by the legislator (Judgment Part III.3.4 p. 30). In effect, life is afforded top priority in the hierarchy of values protected by the law and the life of a foetus (*nasciturus*) is subsumed under the general sense of

---

[6] Judge Pszczółkowski's dissent, introductory section at page 70; translation provided after Gliszczyńska-Grabias & Sadurski 2021.
[7] Judge Pszczółkowski's dissent, introductory section at page 70; translation provided after Gliszczyńska-Grabias & Sadurski 2020.

the concept of *life* under Article 38. In addition, the Tribunal expresses its preference for adopting the *in dubio pro vita humana* interpretative method in the lawmaking process so that any doubts regarding the protection of human life should be resolved in favour of such protection. These two value-laden and ethical concepts of human dignity and life constitute the *topoi* and they are used as underlying premises for the external justification of the court's ruling.

# 6 Weighing and balancing in the Tribunal's reasoning

Weighing and balancing is clearly one of the main argumentative patterns used by the Tribunal in the justification of its decision to tighten the abortion laws in Poland. As already signalled, at the heart of this hard case is a fundamental conflict between constitutional rights and legislation and between competing constitutional rights. We can recall that the majority opinion frames the conflict as occurring between the constitutionally protected value of the life of a human foetus and the legislation permitting abortion in the interests of pregnant women. Balancing as part of a more general principle of proportionality is invariably used in those cases where constitutional powers are exercised and constitutional rights are in competition (Feteris 2017).

In general, balancing is about "determining the priority among competing demands or requirements according to their importance in the concrete case" (Sieckmann 2013, 191). For example, based on a decision given by the German Constitutional Court, Alexy (2003) demonstrates how the collision between the general right to personality (of a convict undergoing his re-socialization) and the broadcasting station's right of freedom of coverage (of a documentary film about a criminal case) can be resolved by weighing and balancing. In that case, the court concluded that the protection of the right to personality should take precedence over the right of freedom of expression (B VerfGE 35, 202 (219). Weighing and balancing involves determining the importance of the principles relevant to the case and judging whether the importance of satisfying one principle justifies the detriment or non-satisfaction of the other (Feteris 2017). Weighing and balancing has been famously expressed in terms of a rule known as the Law of Balancing which is formulated as follows: "The greater the degree of non-satisfaction of, or detriment to, one principle, the greater the importance of satisfying the other" (Alexy 2003, 136). The reasoning in such cases usually consists of three steps. The first step involves assessing the intensity of the interference with the right. Step 2 focuses on evaluating the importance of satisfying the competing principle and Step 3 consists in comparing the intensity of the interference with the right and the importance of satisfying the competing principle in order to conclude whether the interference is at all justified. Two points are worth pointing out about the Law of Balancing. First, despite expressing the formal structure of balancing as a Weight Formula, the procedure is not deductive. Rather, subjective judgments remain crucial elements in this type of reasoning.[8] Second, weighing and balancing presupposes some degree of symmetry and reciprocity in the way the competing principles or rights are considered.

Article 31(3) of the Constitution provides rules to deal with restrictions on constitutional freedoms and rights.[9] Section 4.1 of the majority reasoning refers to the principle of proportionality *sensu largo* and its three elements which, combined, form the so- called 'proportionality test'. The first element is *usefulness rule* which means that restricting a constitutional right can only be done by means of legal measures which result in achieving a goal that justifies the restriction. The second element refers to *necessity rule* according to which, if there are a few means of restricting a constitutional right, because of a constitutionally justified goal, the least oppressive one should be chosen. The proportionality rule *sensu stricto* is the third element of the test. It involves weighing two or more colliding principles and determining which should have a priority in a given legal and factual context. The Tribunal uses the test to rule that the challenged legislation does not meet the criteria for restricting a constitutional right provided in Art. 31(3). In keeping with the general rule of the Law of Balancing, the proportionality test, especially its two first elements, is used to assess the intensity of the interference with the right to life and the importance of the infringement of this principle. The Tribunal asserts that the principle of protecting human life cannot be restricted or infringed "in order to protect values which occupy lower positions in the constitutional [axiological] hierarchy, such as property rights, public morality or environmental protection or even the health of other people." (K 1/20, Section 4.2).

---

8 The extent to which weighing and balancing is a method of rational justification has been the object of some debate which goes beyond the remit of this chapter (see, however, Sieckmann 2013 for a brief overview of the discussion and his solution to this question).
9 Art. 31(3) reads: "Any limitation upon the exercise of constitutional freedoms and rights may be imposed only by statute, and only when necessary in a democratic state for the protection of its security or public order, or to protect the natural environment, health or public morals, or the freedoms and rights of other persons. Such limitations shall not violate the essence of freedoms and rights" (official translation available at https://www.sejm.gov.pl/prawo/konst/angielski/kon1.htm. Last viewed 24th February 2023).

The crucial element of the Law of Balancing, which concerns whether the importance of satisfying the colliding principle and its application "justifies the detriment to or non-satisfaction of the former principle" (Feteris 2017, 146), can be found in the following passage from the majority reasoning (K 1/20, Sec. 4.2):

> In the opinion of the Tribunal, Art. 4a Sec. 1 Item 2 of the Act does not enable one to assume that the high probability of a severe and irreversible foetal impairment or an incurable life-threatening disease is to constitute a basis for the automatic presumption of a violation of the well-being of a pregnant woman, and the mere indication of a child's potential burden concerning such defects is eugenic in nature. This provision lacks any reference to measurable criteria of violating the mother's welfare justifying the termination of pregnancy, i.e. a situation in which she could not be legally required to sacrifice her legal interest.

As can be seen, the reasoning contained in the majority opinion appears to be flawed in that it fails to adequately establish the degree of non-satisfaction of the right to women's self-determination or the women's welfare. Instead, it is preoccupied with establishing the importance of the principle of protecting life and the importance of applying this principle. The Tribunal's reasoning is oriented towards determining the abstract weight of the constitutional right to life that can overrule or 'trump' any other considerations. This constitutional right value, extended to apply to the protection of the life of a foetus is established as a legitimate aim in justifying the adopted legal measure, i.e., restricting the scope of the contested legislative provision. The bulk of the Tribunal's reasoning goes into establishing the importance of this aim in order to justify the infringement on the women's right to self-determination.

It seems that this type of reasoning could be also accounted for by referring to the Law of Trumping, which represents a further development of the theory of balancing (Feteris 2017). Klatt and Meister's (2012) development of the theory focuses on the abstract weights of the conflicting principles or considerations and their effects. In doing so, they formulated the Law of Trumping (Klatt and Meister 2012, 690) which reads: "The higher the abstract weight of a right, the more likely it will trump competing considerations". The Tribunal's reasoning led to the recognition of human life in its prenatal phase as a constitutionally protected value because only constitutional values can constitute a legitimate goal that justifies overruling other considerations.

# 7 Arguments referring to authority: Closing the jurisprudential horizon?

Apart from weighing and balancing, there are other argumentative devices adopted in the Tribunal's reasoning which are worth discussing in order to shed more light on the Tribunal's position and argumentative stance and the way the judges attempt to make their decision acceptable for their audiences. One argumentative practice that judges can apply is *argument ad auctoritatem,* i.e. arguments appealing to an authority or knowledge external to the court or tribunal. References to external sources of authority usually occur when courts feel unable to decide on a matter that may be too specialist for them. It would be perhaps natural to expect the Tribunal to make frequent recourse to this argumentative practice in a case concerning extremely sensitive ethical and medical issues.

Even though case law is not officially considered a source of law in Polish constitutionalism, the Tribunal often refers to its own previous cases, especially the K26/96 case of 28 May 1997, which restricted the admissibility of terminating pregnancy on the grounds of a woman's difficult living conditions or a difficult personal situation. Citing verbatim entire passages from the 1997 case, the Tribunal confirmed the binding nature of its ruling arguing that the axiological aspects determined for that case remain valid for the present case: "What remains constant are two matters: the status of the human foetus and the limits of its legal protection. These matters are treated by the Tribunal as binding and not conditional (K26/96)"

Parallel to the (over)reliance on own case law is the almost complete absence of any reference to the jurisprudence of international courts, especially the European Court of Human Rights. The ECHR has been adjudicating in cases involving the admissibility of terminating pregnancy for almost 50 years and Poland is a party to several recent cases[10] so referring to its caselaw would be perfectly justified. This absence can be viewed as failure to engage with the international human rights framework. Bucholc (2022, 76) uses the concept of jurisprudential horizon to "trace the dynamics of broadening or narrowing down what counts as an eligible reference in jurisprudential practice". This concept offers a way of assessing the choices a court makes when drawing upon interpretations and arguments which are then used in its justifications. While such choices are clearly linked to a given legal culture and the institutional context, they tend to signal a

---

10 See, for example, *Tysiąc v. Poland,* app. no 5410/03, ECtHR judgment of 20 March 2007, R. R. v. Poland, appl. no. 27617/04, ECtHR judgment of 26 May 2011 or P. and S. v. Poland, appl. no. 57375/08, ECtHR judgment of 30 October 2012.

court's changing attitude or argumentative stances. Courts can either broaden their jurisprudential horizons by embracing new arguments or narrow them down by excluding or ignoring them. It should be noted that the expressions 'broadening' or 'narrowing' do not need to be evaluative and, broadening jurisprudential horizons might not always lead to positive effects just as narrowing them should not always be viewed as negative. Rather, it seems more important to determine what is included or excluded as a valid argument from judicial justification. In the context of protecting human rights, courts resolving conflicts between the protection of life and the right to self-determination could be expected to embrace a process whereby "international law and case law have entered the domain of national jurisprudential practice as eligible argumentative references to new sources of interpretation patterns, legitimation, and justification." (Bucholc 2022:76). But this trend cannot be found in the Tribunal's reasoning. There is indeed some evidence that the Tribunal's ruling and its justification could in fact have indicated a trend towards further narrowing down of the jurisprudential horizon regarding human rights.[11]

If the Tribunal made references to external sources of authority, they were of non-juristic nature and concerned the concept of "liberal eugenics". (K 1/20, Section 2.3) The application for the constitutional review in this case posed a question implying that the challenged legislative provision allowing abortion in cases of severe and irreversible foetal abnormalities (Art. 4 Sec. 1 of *the Act on Family Planning*) effectively "legalizes eugenic practices" in respect of unborn children. Without questioning its constitutional relevance and citing several contemporary philosophers, including Francis Fukuyama, Michael Sandel, and Jürgen Habermas, the Tribunal explicitly acknowledges this premise (K 1/20, Sec. 2.3).

The *argument ad auctoritatem* is combined here with the use of a term which evokes clearly negative connotations.[12] On the rhetorical level, the occurrence of the term "eugenics", especially in the phrase "liberal eugenics" in the Tribunal's justification can be interpreted as yet another instance of using an emotive term as an instrument to frame reality (Schiappa 2003) by distorting and redefining its qualities. A situation where prenatal tests or other medical results indicate a high probability of a serious and irreversible impairment of the fetus or an incurable

---

[11] See the decision of the Constitutional Tribunal of 24 November 2021 (K 6/21) that ruled Art. 6 of the European Convention on Human Rights to be unconstitutional.

[12] The negative polarity of the Polish noun 'eugenika' (Eng. eugenics), as well as its adjectival forms 'eugeniczny' and 'eugeniczna' (Eng. eugenic) is corroborated by the analysis of their co-texts in the National Corpus of Polish (http://nkjp.pl/index.php?page=0&lang=1) visited 22.08. 2022.

life-threatening illness of the fetus is assigned a stigmatising label *eugenic* rather than treating it as an embryopathological exception.

# 8 Pulling the strands together

In drawing together the various strands we can see the usefulness and relevance of combining the legal argumentation theory with certain elements of legal rhetoric and persuasion in order to examine judicial justification at different levels. According to MacCormick (2005), justification can be expressed at the internal, deductive and logical level and the external level, which justifies why specific norms derived from a system of values have been selected. This level of justification is external because it often draws upon premises or values that are derived from outside the legal system, i.e. extra-legal values (cf. Leszczyński 2020). This level of justification was the main object of scrutiny in this chapter. The Theory of Balancing was adopted because of its descriptive dimension which allowed me to give a reconstruction of the Tribunal's reasoning that started from arguments expressed in the discourse of legal justification. The weighing and balancing in the majority opinion proceeded with premises that were identified as underlying this process: *dignity* and *right to life*. A critique of how the majority applied weighing and balancing as a method for legal justification points towards a fundamental lack of symmetry in how the colliding principles are considered. A respect for women as human beings should have been weighed against the other protected constitutional values and rights.

The analysis supports the view that argumentation contained in the opinion is unique since it needs to draw on both interpretive as well as rhetorical methods far more often than normally would be the case, in order to explain and justify the Tribunal's controversial decision. The analysis has also revealed that arguments can be used in parallel. For example, the Tribunal refers to external authority to acknowledge the relevance of the negatively charged concept of 'liberal eugenics' to its discussion. In doing so, it relies on value-laden language likely to stir up negative emotive responses. The use of this 'double' argumentative device may have served the purpose of aligning with views of the conservative and illiberal audience which initiated and supported the application for the constitutional review of the abortion law.

The opinion is also noteworthy for its use of evaluative language. Value-laden lexical items occur at several crucial points of the reasoning. They are used as an expression of a specific argument, as in the example of 'liberal eugenics' but they are in fact found both in the majority and dissenting opinions serving the same chief function, i.e. to persuade. In the majority opinion, they denote *topoi*, which provide starting points for the ensuing argumentation. The choice of the specific starting points carries important implications for the argumentative strategies used by the Tribunal. The justification rests on the abstract, high-level concepts of *life* and *human dignity*. From the perspective of acceptability and persuasive potential, this strategy turns out to be dubious and double-edged. It is easy to imagine that audiences opposed to the Tribunal's stance on abortion could embrace the same principles that are used in the court's reasoning but in favour of a different solution. This can be noticed in the dissent, where the same values of life and dignity are invoked in defence of women's dignity and welfare.

Apart from the internal and external levels of legal justification, there is also a wider layer of the political, legal and social contexts in which the ruling is embedded and which compel a different reading and interpreting of the Tribunal's decision. No matter how well, in juridical terms, the justification could be crafted, it would never be effective because of the court's tarnished reputation and a failure to convince its audiences. Romano and Curry (2020, 9) make an interesting distinction between *justification* and *persuasion*. The language of justification is used to claim that the decision reached is the only correct legal solution to resolve a conflict and arguments are selected solely to demonstrate support for it and prove that no other viable choice can be found. In contrast, the language of persuasion is different in that it relies on multiple arguments recognizing the existence of alternatives. It is dialogic in its attempt to show the superior merit of its argument by comparing it with others. Seen from this perspective, the Constitutional Tribunal majority opinion should be viewed as a prime example of justification.

# Bibliography

Aarnio, Aulis. 1987. *The Rational as Reasonable. A Treatise of Legal Justification*, Dodrecht, Reidel.
Afonso da Silva, Virgílio. 2011. *Comparing the Incommensurable: Constitutional Principles, Balancing and Rational Decision*, Oxford Journal of Legal Studies 31(2), 273–301.
Alexy, Robert. 1989. *A Theory of Legal Argumentation. The Theory of Rational Discourse as Theory of Legal Justification*, Oxford, Clarendon Press.
Alexy, Robert. 2002. *A Theory of Constitutional Rights* (translation by J. Rivers of *Theorie der Grundrechte* 1985), Oxford, Oxford University Press.
Alexy, Robert. 2003. *On Balancing and Subsumption. A Structural Comparison*, Ratio Juris 16, 433–449. DOI 10.1046/j.0952-1917.2003.00244.x.
Berger, Linda L. & Kathryn M. Stanchi. 2018. *Legal Persuasion. A Rhetorical Approach to the Science*. London and New York, Routledge.
Bucholc, Marta. 2022. *Abortion Law and Human Rights in Poland: The Closing of the Jurisprudential Horizon*, Hague Journal on the Rule of Law 14,73–99.

Clark, Sherman. 2003. *The Character of Persuasion*, Ave Maria Law Review 1, 61–79.
Feteris, Eveline T. 2017. *Fundamentals of Legal Argumentation. A Survey of Theories on the Justification of Judicial Decisions*, Dordrecht, Springer.
Feteris, Eveline T. & Harm Kloosterhuis. 2013. *Law and Argumentation Theory: Theoretical Approaches to Legal Justification*. Available at SSRN: https://ssrn.com/abstract=2283092 or http://dx.doi.org/10.2139/ssrn.2283092.
Frost, Michael. 2016. *Introduction to Classical Legal Rhetoric*, London/New York, Routledge.
Gliszczyńska-Grabias, Aleksandra & Wojciech Sadurski. 2021. *The Judgment That Wasn't (But Which Nearly Brought Poland to a Standstill) 'Judgment' of the Polish Constitutional Tribunal of 22 October 2020, K1/20*. European Constitutional Law Review 17, 130–153.
Hage, Jaap. 2013. *Construction or reconstruction? On the function of argumentation in the law*. In Christian Dahlman & Eveline Feteris (eds.), *Legal argumentation: Cross-disciplinary perspectives*, New York/London, Springer, 125–144.
Klatt, Matthias & Moritz Meister. 2012. *Proportionality – a benefit to human rights? Remarks on the ICON controversy*. Journal of Constitutional Law 10(3), 687–708.
Koncewicz, Tomasz. 2018. *The Capture of the Polish Constitutional Tribunal and Beyond: Of Institution(s), Fidelities and the Rule of Law in Flux*, Review of Central and East European Law 43(2), 116–173. doi: https://doi.org/10.1163/15730352-04302002
Kovalčík, Michal. 2022. *The instrumental abuse of constitutional courts: how populists can use constitutional courts against the opposition*, The International Journal of Human Rights, 26(7), 1160–1180, DOI: 10.1080/13642987.2022.2108017.
Leszczyński, Leszek. 2020. *Extra-Legal Values in Judicial Interpretation of Law: A Model Reasoning and Few Examples*. International Journal for the Semiotics of Law 33, 1073–1087, https://doi.org/10.1007/s11196-020-09773-y.
Macagno, Fabrizio & Douglas Walton. 2014. *Emotive Language in Argumentation*, Cambridge, Cambridge University Press.
MacCormick, Neil. 2005. *Rhetoric and the Rule of Law. A Theory of Legal Reasoning*, Oxford, Oxford University Press.
Makau, Josina M. 1984. *The Supreme Court and reasonableness*, Quarterly Journal of Speech 70, 379–396.
Perelman, Chaïm & Lucie Olbrechts-Tyteca. 1969. *The New Rhetoric. A Treatise on Argumentation*, Notre Dame, University of Notre Dame.
Romano, Michael & Todd Curry. 2020. *Creating the Law. State Supreme Court Opinions and the Effect of Audiences*, New York and London, Routledge.
Sadurski, Wojciech. 2019a. *Poland's Constitutional Breakdown*, Oxford, Oxford University Press.
Sadurski, Wojciech. 2019b. *Polish Constitutional Tribunal Under PiS: From an Activist Court to a Paralysed Tribunal, to a Governmental Enabler*. Hague J Rule Law 11, 63–84. https://doi.org/10.1007/s40803-018-0078-1.
Schiappa, Edward. 2003. *Defining Reality. Definitions and the Politics of Meaning*. Carbondale/Edwardsville, Southern Illinois University Press.
Sieckmann, Jan. 2013. *Is balancing a method of rational justification sui generis?* In Christian Dahlman & Eveline Feteris (eds.), *Legal Argumentation Theory: Cross-Disciplinary Perspectives*, Dordrecht, Springer, 189–206.
Toulmin, Stephen. 1958. *The Uses of Argument*, Cambridge, Cambridge University Press.

Wigura Karolina & Jarosław Kuisz. 2020. *Poland's abortion ban is a cynical attempt to exploit religion by a failing leader*. The Guardian, 28 October 2020, available at www.theguardian.com/commentis free/2020/oct/28/poland-abortion-ban-kaczynski-catholic-church-protests), visited 22 August 2022.

## Case law cited

B VerfGE 35, 202 (219)

Wyrok Trybunału Konstytucyjnego [Ruling of Constitutional Tribunal] of 28 May 1997, K 26/96, OTK 1997/2/19.

Wyrok Trybunału Konstytucyjnego [Ruling of Constitutional Tribunal] of] of 30 September 2008, K 44/07, OTK-A 2008/7/126.

Wyrok Trybunału Konstytucyjnego [Ruling of Constitutional Tribunal] of 22 October 2020, K 1/20, Dz.U.2021.175.

Wyrok Trybunału Konstytucyjnego [Ruling of Constitutional Tribunal] of 24 November 2021, K 6/21, Dz.U.2021.2161.

## Statutory law cited

Ustawa z dnia 7 stycznia 1993 o planowaniu rodziny, ochronie płodu ludzkiego i warunkach dopuszczalności przerywania ciąży [Act on Family Planning, Protection of Human Foetus and the Conditions to the Admissibility of Pregnancy Termination of January 7, 1993] (Official Journal of the Republic of Poland 1993 Vol. 17, Item 78).

Ustawa z dnia 30 sierpnia 1996 r. o zmianie ustawy o planowaniu rodziny, ochronie płodu ludzkiego i warunkach dopuszczalności przerywania ciąży oraz o zmianie niektórych innych ustaw [Act on the Amendment of the Act on Family Planning, Protection of Human Foetus and the Conditions to the Admissibility of Pregnancy Termination of January 7, 1993 and Amendment of other Acts of August 30, 1996] (Official Journal of the Republic of Poland 1996 Vol. 139 Item 646).

Giovanni Tuzet
# The pragmatics of evidence discourse: Ostensive acts

**Abstract:** The process of juridical proof typically requires some ostensive act, consisting in the presentation of evidence to prove a claim. Building on previous work on the pragmatics of evidence discourse, this contribution concerns the dynamics of ostensive acts and has the purpose of clarifying how such acts work in the context of legal fact-finding, focusing on the use of indexical words in particular. The work concludes by claiming that ostensive acts are necessary but not sufficient to legal fact-finding, for ostension must be followed by argument.

**Keywords:** Acquaintance, Argumentation, Evidence, Indexicality, Legal Fact-finding, Ostension

## 1 Introduction

The theory of legal argumentation traditionally focuses on the arguments given by decision-makers to justify their decisions. Some arguments concern the interpretation of legal provisions (Walton et al. 2020), other arguments concern the evaluation of the available evidence (Anderson et al. 2005). These arguments are typically found in the written opinions of judges. There is less attention, in the literature, to the arguments that parties advance *in vivo*, that is at trial or at some point of the proceedings. These arguments present some particular features that are worth studying.

One feature of such arguments is that they are usually presented orally, in the interaction between participants. What I proposed to call the "Pragmatics of Evidence Discourse" (Tuzet 2021a) is the study of such arguments when related to evidence (cf. Tiersma/Solan 2012, part VI). The parties argue, before the decision-makers, about the evidence and the inferences they want to draw from it.

As a second particular feature, when related to evidence such arguments often refer to something that is present in context. "Presentation of evidence" is indeed a technical phrase in law. Parties present evidentiary items to fact-finders (judges or jurors) by exhibiting things, showing pictures, displaying data, introducing witnesses, etc. Fact-finders are supposed to look at what is shown, to hear

---

**Giovanni Tuzet,** Bocconi University, e-mail: giovanni.tuzet@unibocconi.it

https://doi.org/10.1515/9783110799651-004

testimonies and the like, and to draw the appropriate inferences. These inferences are often suggested by using arguments from ostension as "invitation to inference" (Marraud 2018), that is, by showing the relevant items to the fact-finders and suggesting how to inferentially process them. The use of indexicals and demonstratives, as belonging to deixis more generally speaking (Levinson 1983, ch. 2), is instructive in this respect.

In those legal contexts, ostension is not performed for definitional or conceptual purposes (as "ostensive definitions" are), but for probatory purposes. The litigated facts can be proven by presenting evidence that supports the relevant factual claims. This is not just a possibility: the process of juridical proof typically requires some ostensive act, consisting in the presentation of evidence to prove the relevant claims. In this sense, evidentiary items are elements which are susceptible of being shown, or exhibited, or indicated by the parties to the fact-finders in the relevant context.

However, evidence by itself does not prove anything. One has to construct evidentiary arguments, from the evidence presented to the probatory conclusions that one "invites" to draw. *Evidence without inference is blind and inference without evidence is void.* This rephrasing of the Kantian motto[1] means that, on the one hand, ostensive acts need arguments about their probatory consequences and that, on the other hand, probatory arguments need the ostension of evidence.

The present work is about the dynamics of ostensive acts and has the purpose of clarifying how such acts work in the context of legal fact-finding. It builds on previous work on the pragmatics of evidence discourse.

Section 2 recaps what I found in earlier work on the modalities of ostension and the logic of ostensive acts (Tuzet 2022). One such modality is the use of indexical words. Section 3 deals with indexicality in law and philosophy. Section 4 deals with the need of drawing arguments from the ostended evidence. Section 5 concludes by stressing that ostensive acts are necessary but not sufficient for legal fact-finding. Future research is needed on the logic of arguments from ostension.

## 2 Probatory ostension

In order to move from evidence to legal verdict five requirements at least must be satisfied: (1) evidence must be admissible according to the rules of the relevant legal system; (2) evidence must be presented to fact-finders; (3) evidence must be "inferentialized" by parties and fact-finders, since evidence does not speak for itself

---

**1** According to it, thoughts without content are empty and intuitions without concepts are blind.

and the participants to a dispute have to construct evidentiary arguments based on the items presented; (4) evidence must be assessed to determine its probative value (or, better, evidentiary inferences and arguments must be assessed to determine the evidential support, or warrant, or justification provided by premises to conclusions); (5) fact-finders need to consider whether the evidence (as assessed) meets the relevant standard of proof, or, to put it differently, whether the relevant burden of proof has been satisfied.[2]

This Section of the present paper is on the second requirement of the above list. It summarizes how ostensive acts work in the context of legal fact-finding.

As a preliminary point, we must distinguish the probatory use of ostension from its definitional use. Probatory ostensions are not ostensive definitions. As the name makes it clear, the act of ostensive definition consists in showing some item that indicates the meaning of a word or expression. For instance, showing a yellow thing in order to indicate what we mean by "yellow", or pointing at an English horn in order to show what we mean by "English horn". Examples easily multiply.

To cut a long story short, *ostensive definitions* were cherished by the logical positivists of the Vienna Circle, who tried to use them to explain how our words and concepts connect to the world of empirically detectable things (Coffa 1991, 176–177, 242, 249ss. and 354ss.; Misak 1995, 89–96). But they were also seen with suspicion and criticized by Wittgenstein (1953, §28ss.), who claimed that no ostensive act can convey any meaning unless it is part of an already established language game.[3] Quinean speculations on "Gavagai" and the indeterminacy of translation fueled reservations on ostensive definition: when natives say "Gavagai" in presence of a rabbit, do they refer to a rabbit, to food, to a momentary rabbit-stage, to an undetached rabbit-part? (Quine 1960, 29ss.). However, recent work on the operations of our cognitive system seems to vindicate the role of ostensive definition.[4]

---

[2] On this last point, see Allen 2014. For a sophisticated account see also Nance 2016.

[3] "So one might say: the ostensive definition explains the use – the meaning – of the word when the overall role of the word in language is clear. Thus if I know that someone means to explain a color-word to me the ostensive definition 'That is called 'sepia'' will help me to understand the word . . . One has already to know (or be able to do) something in order to be capable of asking a thing's name. But what does one have to know?" (Wittgenstein 1953, §30).

[4] Engelland (2014, 174–175) claims that ostensive definition is helped by the following factors: (a) bias toward the novel (we naturally attend to what is novel in experience); (b) perspectival bias toward an object of a certain size in the perceptual field; (c) bias toward a certain kind of thing (like basic actions and present things); (d) bias toward essential properties (we look for them, not for accidental ones, absent indicators to the contrary); (e) bias toward a certain level of generalization (some low level of generalization, for we don't define particulars, nor very abstract terms); (f) conversational context (which can reverse or specify the above biases).

In contrast, *probatory ostensions* do not define terms or expressions. We perform them to prove a disputed fact. More technically speaking, it is the use of ostension to support a factual claim when the claim is controversial, pointing at the evidence that justifies to some extent the disputed claim.

If one has theoretical ambitions, it is possible to turn that into a conceptual account of juridical evidence. The process of juridical proof typically (or necessarily?) requires some ostensive act. In this sense juridical evidence can be taken to consist in such elements that are *not only perceptible by those who are present in context, but also susceptible of being shown, or exhibited, or indicated to the fact-finders in the relevant context*. Here I will put those theoretical ambitions to one side. My only purpose is to make some clarifications about the kind of ostensive acts that are required and performed in legal fact-finding.

Now two points deserve particular attention: the variety of ostensive modalities, and the logical structure of ostensive acts in the context of legal fact-finding. The former leads to a functional conception of ostension.

## 2.1 A functional conception of ostension

Consider pointing with one's finger, pointing with the glance of the eye, or a motion of the head, or a vocal gesture, and so on. These are ostensive modalities. Ostensive acts can be performed with any of the modalities that are sufficient to indicate something, and sufficient to draw someone else's attention to it. This leads to a functional conception of ostension, where an act counts as ostensive if it has the function of showing something in the relevant context.

In some contexts a gesture can suffice and be better than a nod. In some others a couple of words can be adequate (e.g. "Here's the document"). In some situations it is good to employ more than one modality at the same time (e.g. a gesture and a presenting word or expression). Situations differ in these respects.

In legal proceedings rarely a gesture suffices. True, in a lineup identification procedure a pointed finger may suffice. But that is a kind of peculiar situation. In the central case of ostension in legal proceedings the act is accompanied by words or statements uttered by the relevant party to the relevant hearer. A physical object is shown to the jury and the prosecutor explains that it was found in the defendant's apartment; a picture is shown to the jury and the defendant explains that it was taken just after the crime; a witness is introduced and questioned; etc.

If we focus on the linguistic modalities of ostension, we can realize that speech acts about juridical evidence have some particular features. Probatory discourse is characterized by the use of indexicals and demonstratives much more than normative discourse is.

As an established convention in contemporary pragmatics, we call *indexicals* such words as "I", "you", "here", "now". These are words that refer to something without naming it, nor giving a description of it. And we call *demonstratives* the words, like "this" and "that", that perform the same basic function by pointing at something.[5] We need not discuss here their distinction and the properties that identify each of the two categories. For simplicity's sake, let us use "indexicality" in a broad sense encompassing the use of demonstratives and of indexicals properly understood. Section 3 will explore below some interesting features of indexicality in this broad sense and with respect to legal fact-finding.

## 2.2 The logical structure of ostensive acts

What is the logical structure of an ostensive act in legal proceedings and in a trial in particular? We can see it by distinguishing its logically essential components.

Essentially, an ostensive act has this structure: $A$ shows $B$ to $C$ in context $D$ to draw inference $E$. For what matters here, $A$ is the party that exhibits the evidence. $B$ is the evidentiary item. $C$ is the fact-finder to whom the evidence is shown. $D$ are the spatiotemporal coordinates defined by law, namely the relevant legal context. And $E$ is the inference that the fact-finder is expected to draw given the evidence that has been shown.

Some comment is in order on $D$. The relevant context, for instance a trial or a hearing, is defined by the law. Legal rules – procedural and evidentiary ones – establish the relevant spatiotemporal coordinates. Where, when, how, by whom and to whom the evidence can be shown. This is a peculiar aspect of law. It makes legal fact-finding significantly different from ordinary cognition and from scientific inquiry, which in general is conducted without normative delimitations and constraints. Scientists can choose the subject-matter of their inquiry, the ways to perform it, the amount of time they devote to it, the kind of evidence they look for, and so on. Parties and fact-finders, instead, play their roles in a context which is highly regulated by the law. Players are bound by the rules in several respects, in-

---

[5] According to the received view, the difference between the two categories is that only indexicals enjoy semantic rules which determine their reference (for instance, "I" refers to the speaker). See Levinson 1983, 54ss.

cluding spatiotemporal limitations, bans on certain kinds of evidence, legal criteria of evidence assessment, and mandatory standards of proof.[6]

Additionally, note that the above structure better fits certain kinds of proceedings than others. Ostensive acts take place when those who make decisions are not the same legal actors who collect the evidence. This division of labor typically occurs in adversary proceedings where litigating parties, making their claims, present the evidence they have to independent decision-makers. It does not occur in so-called inquisitorial proceedings, where the judge also acts as a prosecutor and can even use anonymous testimonies that are not presented in court. In inquisitorial contexts the same actor plays both roles, namely collecting the evidence and deciding the case; then the need of ostension evaporates. Public ostensive acts take place before independent (and supposedly impartial) fact-finders. The purely adversarial model of adjudication is the one that maximizes the importance of probatory ostension.

This rather technical point signals a more important aspect. As Twining (1984, 267) pointed out, evidence is "the means of proving or disproving facts". Typically, what one party is interested in proving, the counterparty is interested in disproving. For example, when the plaintiff tries to prove negligence in a civil tort case, the defendant is interested in disproving negligence. Hence, ostensive acts, especially in adversary contexts, can either have the purpose of supporting a factual claim or the purpose of attacking it (by showing it is inconsistent, inadequate, inaccurate, etc.). The tendency to think that juridical evidence is collected and exhibited only for verification or confirmation purposes is an instance of the so-called confirmation bias (Kahneman et al. 1982, 149–150). Inference $E$ "invited" in context $D$ by $A$ showing $B$ to $C$ can be an inference whose conclusion is the falsification of the claimant's factual standpoint. But of course, it can also be the inference of the claimant's conclusion given the evidence. Or an additional point made by the claimant in resisting the counterparty's attempt to falsify the principal standpoint. Ostensive acts are part of this dialectical dimension.

## 3 Indexicality

Evidence discourse as performed by the parties in a dispute is likely to be significantly characterized by the use of demonstratives and indexicals. Speech acts in this context are frequently indexical. They are ostensive acts. "Here is the object

---

[6] See e.g. Haack 2004; Goldman 2005; Bulygin 2015, 219, 257–261. The usual phrase for rules that exclude some kinds of evidence (especially illegally obtained one) is "exclusionary rules".

that was found in the apartment"; "That is the picture which was taken just after the event"; "This is the most important witness"; "Listen to this declaration"; "Look at that message": these are simple examples of what we are considering, that is, indexical evidentiary statements made with the ostension of the relevant item. Similarly for the linguistic interaction with witnesses: "Did you see her?"; "That is the person I saw run away from the bank"; "Do you recognize this voice?"; "This is John's voice, Paul's is that", etc.

An interesting feature of indexicality from a philosophical viewpoint is the connection which it establishes with things that belong to the real world. You can point at real, existing things. You cannot point at a building that has been demolished – but of course you can point at an existing image of the thing.[7] The use of indexical words is a linguistic way to perform the ostensive function outlined above. Charles Sanders Peirce pointed out that feature and discussed the "existential connection" that indexicals have with their reference.[8] Among others, Hilary Putnam (1975, 271) has elaborated on this and claimed that ignoring the indexicality of most words is ignoring the contribution of the environment to our language and cognition.[9] On a different note, Willard Van Orman Quine (1960, 100) has highlighted the utility of demonstrative terms: first, by using expressions like "this river" we are saved the burden of knowing names (when objects have them); second, we can refer to objects that have no proper names ("this apple"); third, we are aided in the teaching of proper names. The economy of effort is remarkable. A demonstrative like "this" serves also as a singular term by itself, and can be used with mass terms ("this water") (Quine 1960, 101). The literature on those topics is abundant.

Let us pause on Peirce in particular. He called "indices" those signs (including words but not limited to them) that perform an indexical function (Peirce 1998, 5, 8, 274, 291). A pointing finger is the standard example. Peirce's scholars have added to this. "Instinct, custom or convention draws our attention to the extended arm and rigid finger and tells us to look along the line it defines . . . Instinct or convention calls attention to the line but does not make that line."[10] For an index

---

7 True, you can also point at ruins or at some stones the building was made of. These are existing things.
8 For an illuminating comment see Burks 1949; see Atkin 2005 for a sophisticated categorization. See also Short 2007, 48ss., 192ss., 214ss.; and Short 2004, 219–222.
9 Cf. Putnam 1975, 229–235. See also Kripke 1980. Consider in addition the indexicality of proper names (of parties, places, etc.) in the trial oral exchange and in the written documents including judicial opinions (case-specific information with proper names is necessary, for an indexicality-free description of the case is impossible); but note that this form of indexicality does not necessarily require ostension.
10 Short 2004, 221. Compare that with Wittgenstein's reservations (on ostensive definition) considered above.

to work, there must be some "spatial connection" between two actual entities (Short 2007, 219), or a "physical contiguity" with the objects referred to (Atkin 2005, 164). However, as commentators have noted, indices are rarely pure: "indices are interpreted in light of other signs with which they occur ... or in light of background knowledge about indices of their type" (Short 2004, 221). The examples given above – "Look at that message", "Listen to this declaration", etc. – instantiate the use of indexicals in combination with other signs. A pure indexical – "That!" – does not tell anything out of very specific contexts.

Additionally, other linguistic signs like "message" or "declaration" bring with them concepts and conceptual assumptions. Assumptions about causation have a special interest here. The fever example is instructive in this respect:

> Higher than normal bodily temperature is called "fever" and is taken to be a symptom of infection only because we know that generally such a condition is caused by infection. Nonetheless, the individual instance of fever picks out an individual instance of infection by virtue of being causally connected to it, and that causal relation is independent of any ideas we may have about it. (Short 2004, 221)

So, cognition combines indices and concepts. Objects must be apprehended conceptually (Short 2007, 51) even when, as natural objects are, they are independent from our concepts. Ignoring your fever does not make it disappear.

Causal relations can be complex. Some signs are the effects of their objects, as a symptom is the effect of a disease. But an ostensive modality like a pointing finger has an intentional component. Compare the relation between dawn and cockcrow and the one between dawn and morning reveille: "In the case of the cockcrow, the object causes the sign and so this is a genuine index. In the case of morning reveille, the object of the sign is, at best, part of a broad causal nexus." (Atkin 2005, 180) I would say that in the former relation there is just efficient causation, whereas in the latter there is final causation.[11]

Notice an important distinction that is implicit in the preceding discussion: there can be an item which is an index of something else, like a symptom is an index of a disease; and there can be something that points at that index, like a finger pointing at the symptom. The latter is a higher-order indexical relation, in that there is an index about an index. When the meta-indexical relation (the index about an index) is instantiated by an ostensive act, there is an element of intentionality and final causation, or teleology, that is absent from purely natural

---

[11] Consider the pointing finger too in this respect: in pointing there is an element of intentionality and final causation. But for Atkin (2005, 181) the causal reading of all indices is quite problematic; as an alternative he suggest an "informational" reading of indices, that is, the idea that a genuine index not only indicates its object, but provides information about it.

processes governed by efficient causation. And in very complex situations there can be indices about indices about indices etc.

Now, objects referred to in ostensive acts can certainly have probative value. In some legal contexts they are called "exhibits" (Allen et al. 2006, 173ss.). Typically, they causally result from the events they are evidence of. But the indexical nexus becomes tenuous when more than one causal chain may produce some such result (for instance, a blood trace can be on the crime scene for several reasons).

Things become increasingly problematic when we consider testimonial evidence such as the one provided by witnesses. "That is the person I saw run away from the bank" is an example. Witnesses can be reliable and sincere, but they can also fabricate their testimony. They are hardly indices of what they say – or they are so in a very loose sense. To put it in Gricean terms (Grice 1957), the meaning of a testimony is a matter of "non-natural" (or conventional) meaning, whereas an object or an image can have a "natural" meaning (being a genuine index). Written documents are similar to testimonies in this respect. Both documents and oral testimonies employ the conventional signs of a language and, to some extent, they need to be interpreted. If a testimonial statement is ambiguous or unclear, interlocutors can ask for clarifications. If a document is ambiguous or unclear, and its drafters are unavailable or disagree, interpreters must supplement its wording and make sense of it in the context of the legal system and its normative requirements. Contracts are often incomplete in that unlikely contingencies are not provided for, or because bargaining about some circumstances would have more costs than expected benefits; then, if some such circumstance becomes relevant in litigation, decision-makers have to supplement the wording of the contract. All of this is to say that ostended evidence (physical, documentary, testimonial) is always necessary but hardly sufficient to legal fact-finding and judicial decision-making.

## 4 Ostensive acts and arguments

Consider a beauty contest. Imagine to be one of the jurors. Would you expect to, and prefer to, (1) have acquaintance with the candidates, or (2) see pictures of them, or (3) hear testimonies about them? Of course, (1) is the usual thing.[12] For the relevant activity purposes, (2) is better than (3) but worse than (1). Seeing the candidates in person gives a better idea of their aesthetical properties. Seeing pic-

---

[12] The beauty context example was suggested by Keynes (1936, 156) for a different purpose, namely giving an account of investments that come from a general second-guessing of other agents' behavior; also, it concerned newspaper photographs of the candidates.

tures can be informative but misleading in some respects. Hearing testimonies can be even worse because, as we pointed out, witnesses can be misleading, inaccurate, unreliable, untruthful, malicious, and the like.[13]

Acquaintance brings us closer to what Locke called the "Original Truth" (*An Essay Concerning Humane Understanding*, book IV, ch. XVI, §10). This is a fundamental reason for preferring the direct inspection of an item over a picture of it; and to prefer a picture over a testimony; and firsthand testimony over hearsay (Tuzet 2021b), other things being equal.[14]

Now, ostensive acts provide acquaintance with the things referred to. In this sense ostensive acts get someone closer to the Original Truth, because they enable direct knowledge of the things referred to. Getting closer to it does not mean hitting it, of course. Peirce, again, gives a valuable lesson for our cognitive life, namely *fallibilism* (Peirce 1998, 42ss.; Haack 2013, 199ss.). Our cognitive efforts are always susceptible to error and failure – which does not mean that we are always wrong (fallibilism is not skepticism). Ostensive acts can be misleading, for they can point at the wrong things, and "invite" the wrong inferences even when they point at the right things. This is the somber part of the story about ostensive acts, which does not cancel their value, though.

In the following I will focus on two ways of overstating the value of ostensive acts. They belong, respectively, to the philosophical and legal discourse.

In philosophy there have been attempts to highlight and vindicate the role of ostension in word learning and language acquisition. Chad Engelland (2014, 194) claims that the logic of ostension is like our own two legs, "because it is intrinsic to what we are as embodied or animate minds." For his purposes, he defines "ostension" as "bodily movement that manifests our engagement with things, whether we wish it or not", and he claims that "ostensive acts make intentions manifest rather than motivate inferences to hidden intentions" (Engelland 2014, xi, xxiii). Note that, in the defined sense, ostension is not necessarily verbal, nor necessarily intentional; and note that, as to ostensive acts, they manifest intentions in action dispensing the observers from making inferences to

---

[13] For a brilliant taxonomy of deception, see Horn 2017.
[14] See Iaquinto/Spolaore 2019 for an outline of a logic of "knowledge of acquaintance". They prefer "knowledge of acquaintance" (James) to Russell's "knowledge by acquaintance" since, as they say, the latter is generally used to denote a form of knowledge-that, i.e. a form of propositional knowledge that does not capture the features of the former (instantiated by examples like "John knows Sarah"). A guiding thought is that *existence* is a necessary condition of knowledge of acquaintance; I would add it is also a necessary condition of ostensive acts.

other minds.[15] This is indeed a very broad definition of "ostension". The typical ostensive act that we have been considering, namely pointing with a finger, has an intentional character and a more specific point, namely the purpose of providing some knowledge or proving a disputed claim.

The idea that behavior reveals intentions regardless of communicative intentions seems to be correct, though. What I am rather doubtful of is the reach of the view that we need not make inferences from behavior to mental states. In some cases it can be a matter of routine, a matter of ordinary circumstances that do not require reflection and reasoning. In other cases behavior is puzzling and we need guesses ("abductions", in Peirce's terminology)[16] to explain it. Suppose a friend behaves differently from the usual way; you wonder why and try to infer the mental states that explain the unusual behavior. Suppose a party to a contractual relationship does not deliver the expected goods; before filing a lawsuit you ask why and try to understand what is going on. When doubts concern people's behavior we can simply ask them, if we have a chance to do it. When they concern sub-personal or supra-personal phenomena we cannot ask (only in a metaphorical sense we "ask" questions to nature by running experiments). And when we ask people why they behave in a certain unexpected and puzzling way we have no guarantee that their answers will be sincere and clear enough. Sometimes we stop asking because we feel it is impolite, sometimes because it is useless. Often we need to interpret answers and cannot go on asking whether our interpretations are correct. Contexts where interests diverge are typical in this respect: collaboration stops and intentions can become strategically opaque.[17] Questioning witnesses in a trial shows these different possibilities: sometimes witnesses collaborate in the most sincere and reliable way, sometimes they do not and can be hostile. Bargaining behavior and political negotiations too, *mutatis mutandis*, show that. Taking behavior and ostensive acts (in Engelland's broad sense) at face value can be a way of overvaluing them.

In philosophy there are also recent views that stress the epistemological and practical value of *gestures*. So-called "gesture studies" (Kendon 2004; McNeill 2005) categorize our individual and social gestures, explore their cultural compo-

---

[15] Note the difference with Sperber and Wilson (1995, 49), claiming that ostension "makes manifest an intention to make something manifest". "When we take ostension more broadly as I do, what ostension manifests is not an intention per se but the target of an intention, namely, a publically available item in the world. Ostension, then is behavior that makes manifest the target of our intentions, and it can occur either with or without an intention to communicate." (Engelland 2014, 31).

[16] See Peirce 1998, 106ss. 205ss., 231ss. The literature on abduction is vast. See e.g. Bellucci 2018, and Tiercelin 2018 (claiming that abduction can be seen as an "epistemic sentiment", which helps overcome the discovery/justification divide).

[17] Game theory addresses these contexts in particular. See e.g. Osborne/Rubinstein 1994.

nents, and point out their value in verbal communication and thought. Philosophers in the pragmatist tradition (Maddalena 2015; Fabbrichesi 2017) add to this by claiming that gestures have the virtue of synthesis (opposed to analysis) and govern the acquisition of knowledge and our meaningful behavior, being actions "with a beginning and an end that carries a meaning" (Maddalena 2015, 69); or by claiming that gestures have "pragmatic unity par excellence" for they consist in meaningful behavior that triggers a social response (Fabbrichesi 2017, 54). The philosophers of gestures allege their role in education, morality, thought and creativity.

Again, that has a broader scope than the one of the ostensive acts we have been discussing with reference to legal proceedings in particular. And that sort of approach to gesture carries the risk of overstating the value of ostensive acts. Let alone insincere and misleading gestures, in controversial contexts a gesture or ostensive act rarely suffices: to settle a dispute or prove a litigated claim we need arguments that build upon ostended evidence. The law is one of such contexts.

Let us turn back to law then. The "Immediacy principle" in Continental jurisprudence is the view, simply put, that fact-finders must have direct knowledge of the evidence and of its sources. In particular, fact-finders must hear the witnesses directly and, at the same time, monitor their behavior in court (what is called, in legal terminology, "witness' demeanor"). This sort of acquaintance with the evidence and its sources is supposed to guarantee a better assessment of it and a more accurate reconstruction of the relevant facts. Ostensive acts permit it. The closer they bring to the Original Truth, the better.

Common law countries have a functional equivalent of the Continental Immediacy principle: it is the trust in the factual determinations of the jury or of the trial judge. More technically speaking, it is a matter of deference. Juries and trial judges hear the testimonies, see the exhibits, get acquainted with the evidence. Appellate judges do not (unless they are given specific powers). This explains and justifies the usual deference of appellate judges in the common law world towards the admissibility decisions of trial judges and the factual determinations of the jury.

So far, so good. The problem is that fact-finders often overvalue the impressions they get from that sort of acquaintance. The risk is to substitute feelings for cognition, and subjective impressions for reasoned decisions. In fact, many scholars have been criticizing over the last years the reach of the Immediacy principle and the value that is put in the direct examination of evidentiary sources.[18] The literature on heuristics and biases encourages this critical standpoint. The "hunch" of the fact-finder risks to be the operation of a bias and as such it cannot justify a

---

[18] See e.g. Ferrer 2007, 62ss.; Gascón 2012, 203ss.; Dei Vecchi 2018, 69ss. All these authors stress the need of a *rational* evaluation of evidence.

decision. However, we should not throw the baby with the bath water. What is wrong is neither the immediacy requirement in itself nor the dynamics of ostensive acts before the fact-finders. What is wrong is the overvaluing of that. More cognitive humility and awareness of our fallible nature would do probably better than a replacement of the immediacy requirement by some paternalistic artificial intelligence system or some algorithm supposedly computing the value of the evidence and giving the human decision-makers a resulting probability measure (let alone the fact that, in some sense, the artificial intelligence system would need an input from evidentiary acts such as introducing and questioning a witness).

In a nutshell, to prove a controversial claim we need indices, concepts and inferences presented in probatory arguments.[19] With a proviso that the following example is a bit extreme, consider the situation where medical pictures are exhibited to the fact-finders:[20] Figure 1 is a brain picture of an allegedly insane person; Figure 2 is a brain picture of a sane one. The white ellipses in both of them are supposed to point at a difference in their brain states. What we immediately see is a difference in color (more green and blue in Figure 1, more red and yellow in Figure 2) and some difference in shape. What can we infer from this? Nothing, as lay persons. Only experts can make inferences and build arguments upon this evidence.

Note that those pictures have an indexical dimension in that they are the effect of actual brain states;[21] additionally, the ellipses have an indexical function in that they draw the observer's attention to some specific features (differences in color and shape). Absent the ellipses, one could point at those features in a different way (e.g. with a laser pointer while projecting the images). Showing the pictures and pointing at those features are ostensive acts. What is their value? They certainly present evidence which is allegedly relevant to the disputed point, namely whether the defendant acted in a way affected by insanity and then should be granted a defense. The question is whether such evidence has the alleged probative value. In order to answer this question one has to build an argument that explains the differences in color and shape and connects them to

---

[19] Note that an inference can be purely mental and private, whereas arguments properly speaking are public as directed to an audience. In a way, inferences are the logical structure of arguments.

[20] I take them from an expert report in an Italian insanity case of 2018. I thank Pietro Pietrini and Ilaria Zampieri for handing them over to me. On expert signs and legal burdens, see Tuzet 2023. See also Hammel 2022 on linguistic expert evidence (e.g. on expert assessment of statements presented as evidence).

[21] Semiotically they also have a dimension that Peirce called "iconic": to some degree, they resemble their objects (see Peirce 1998, 4–10, 267–299). Note in addition that Peirce stressed the role of diagrams that *show* inferential relations.

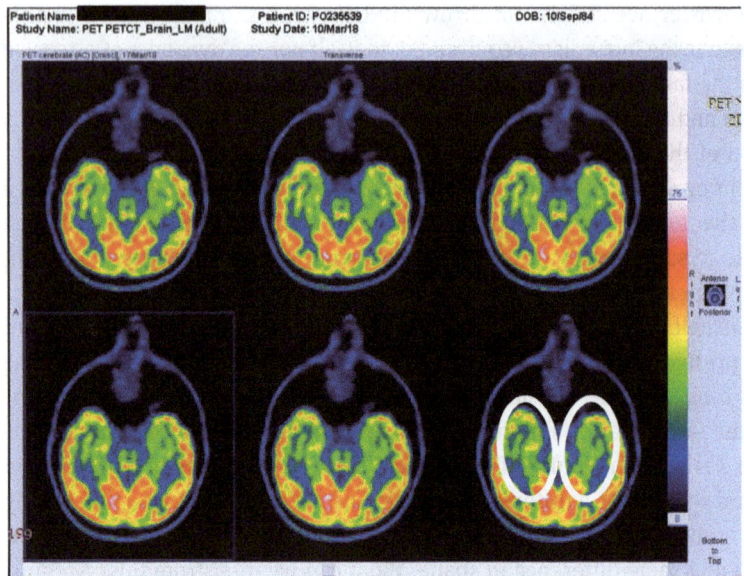

**Figure 1:** (allegedly insane person).

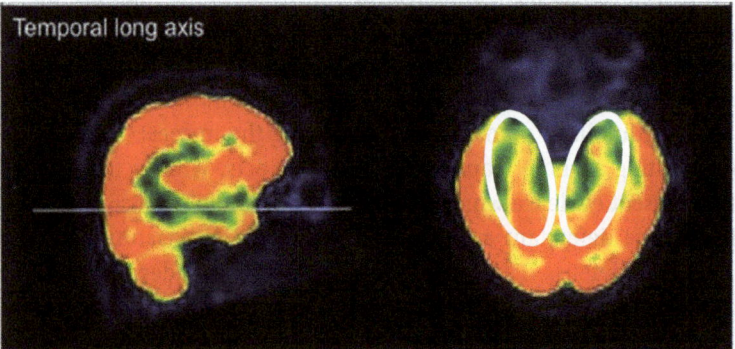

**Figure 2:** (sane person).

actual brain states, insane or not. This, among other things, requires a concept of insanity. And it requires inferential processing. You cannot simply read insanity out of the pictures.[22]

---

[22] In fact the above pictures are part of a 90-page expert report full of technical information and expert data.

To use a phrase that comes from philosophers of science, we need in particular some specific "bridge-laws" (Hempel 1966; Horgan 1978; Tiercelin 2011, 160ss.) that map mental states onto brain ones. "Bridge-laws" are used for several purposes, that go from the theoretical reduction of one sort of the phenomena to the other, to the probatory use of phenomena of one sort to prove phenomena of the other (e.g. proving mental states, dispositions or inabilities, given brain states).

Only experts can perform the conceptual and inferential processing that such indices require. And the additional legal problem is that, in litigation, experts often disagree with one another. Typically, the forensic expert called by the prosecution has a view (say, the person was sane at the time of the act) and the expert called by the defense has the opposite view (the person was insane). Independent experts appointed by judges can help to some extent but are not error-free witnesses. Fact-finders, especially jurors, are often at a loss when experts disagree. The only reasonable solution, in many cases, is to apply the burden of proof rules: if the ostended evidence and the arguments made out of it do not discharge the relevant burden of proof, then the burdened party loses.

The above example is a bit extreme because it looks like peculiar. Not every item of evidence is expert, nor is every ostensive act performed by an expert. Non-expert evidence can be less problematic, true. But "factual ambiguity" (Allen 1994) is a widespread phenomenon in legal fact-finding. Assume a blood trace is found at the crime scene and nobody disputes that the trace comes from the defendant. Is this sufficient to convict? Usually it is not. Absent an expert dispute on the trace itself, the defendant might claim that the trace was left earlier when defendant and victim dined together and a knife cut the defendant's finger; or because the victim assaulted the defendant, who acted out of self-defense; and so on. Given the evidence, there can be more than one explanation compatible with background knowledge and common sense assumptions. Again, ostension must be followed by argument.

# 5 Conclusion

The preceding discussion was meant to show that ostensive acts are necessary but not sufficient to a reasoned decision in legal fact-finding. Ostension must be followed by argument. Arguments often diverge in their understanding and explanation of the available evidence, as well as in the probatory conclusions that they "invite" to draw. When the evidence is testimonial (especially when it comes from expert witnesses) fact-finders have the additional problem of interpreting correctly what the witnesses say. But the point can be generalized to any probatory argument advanced by the parties. Probatory arguments are given orally (or in written form), and a fact-finder must first determine their content and then assess their value. This is not something that naturally flows from ostensive acts.

That should not be taken as a skeptical remark. As cognitive inputs of the fact-finding process, ostensive acts have great importance and cannot be dispensed with. Only, one must be aware of their limits. And we should consider that one can also show a diagram which displays an inferential structure, that is, an argumentative representation of how a certain conclusion follows from the given premises. Showing a diagram can be an ostensive act that indicates how to move from the cognitive inputs to the appropriate inferential output.

# Bibliography

Allen, Ronald J. 1994. *Factual Ambiguity and a Theory of Evidence*, Northwestern University Law Review 88, 604–640.
Allen, Ronald J. 2014. *Burdens of Proof*, Law, Probability and Risk 13, 195–219.
Allen, Ronald J., Richard B. Kuhns, Eleanor Swift & David S. Schwartz. 2006. *Evidence. Text, Problems, and Cases*, 4$^{th}$ ed., New York, Aspen.
Anderson, Terence, David Schum & William Twining. 2005. *Analysis of Evidence*, 2$^{nd}$ ed., Cambridge, Cambridge University Press.
Atkin, Albert. 2005. *Peirce on the Index and Indexical Reference*, Transactions of the Charles S. Peirce Society 41, 161–188.
Bellucci, Francesco. 2018. *Eco and Peirce on Abduction*, European Journal of Pragmatism and American Philosophy 10, 1–20.
Bulygin, Eugenio. 2015. *Essays in Legal Philosophy*, ed. by Carlos Bernal et al., Oxford, Oxford University Press.
Burks, Arthur W. 1949. *Icon, Index, and Symbol*, Philosophy and Phenomenological Research 9, 673–689.
Coffa, J. Alberto. 1991. *The Semantic Tradition from Kant to Carnap. To the Vienna Station*, Cambridge, Cambridge University Press.
Dei Vecchi, Diego. 2018. *Problemas probatorios perennes*, Fontamara, México.
Engelland, Chad. 2014. *Ostension. Word Learning and the Embodied Mind*, Cambridge (Mass.) and London, The MIT Press.
Fabbrichesi, Rossella. 2017. *The Iconic Ground of Gestures. Peirce, Wittgenstein and Foucault*. In Kathleen A. Hull & Richard K. Atkins (eds.), *Peirce on Perception and Reasoning. From Icons to Logic*, London, Routledge, 54–60.
Ferrer, Jordi. 2007. *La valoración racional de la prueba*, Madrid, Marcial Pons.
Gascón, Marina. 2012. *Cuestiones probatorias*, Bogotá, Universidad Externado de Colombia.
Grice, H. Paul. 1957. *Meaning*, The Philosophical Review 66, 377–388.
Goldman, Alvin. 2005. *Evidence*. In Martin P. Golding & William A. Edmundson (eds.), *The Blackwell Guide to the Philosophy of Law and Legal Theory*, Oxford, Blackwell, 163–175.
Haack, Susan. 2004. *Truth and Justice, Inquiry and Advocacy, Science and Law*, Ratio Juris 17, 15–26.

Haack, Susan. 2013. *Putting Philosophy to Work. Essays on Science, Religion, Law, Literature, and Life*, expanded ed., Amherst (New York), Prometheus Books.
Hammel, Andrew. 2022. *Linguistic Expert Evidence in the Common Law*. In Victoria Guillén-Nieto & Dieter Stein (eds.), *Language as Evidence. Doing Forensic Linguistics*, Palgrave Macmillan, Cham, 55–84.
Hempel, Carl G. 1966. *Philosophy of Natural Science*, Englewood Cliffs, Prentice-Hall.
Horgan, Terence E. 1978. *Supervenient Bridge Laws*, Philosophy of Science 45, 227–249.
Horn, Laurence R. 2017. *Telling It Slant: Toward a Taxonomy of Deception*. In Janet Giltrow & Dieter Stein (eds.), *The Pragmatic Turn in Law: Inference and Interpretation in Legal Discourse*, Berlin, De Gruyter, 23–55.
Iaquinto, Samuele & Giuseppe Spolaore. 2019. *Outline of a Logic of Acquaintance*, Analysis 79, 52–61.
Kahneman, Daniel, Paul Slovic & Amos Tversky (eds.). 1982. *Judgment under Uncertainty. Heuristics and Biases*, Cambridge, Cambridge University Press.
Keynes, John Maynard. 1936. *The General Theory of Employment, Interest and Money*, London, Macmillan.
Kendon, Adam. 2004. *Gesture. Visible Action as Utterance*, Cambridge, Cambridge University Press.
Kripke, Saul A. 1980. *Naming and Necessity*, 2$^{nd}$ ed., Oxford, Blackwell.
Levinson, Stephen C. 1983. *Pragmatics*, Cambridge, Cambridge University Press.
Maddalena, Giovanni. 2015. *The Philosophy of Gesture. Completing Pragmatists' Incomplete Revolution*, Montreal and Kingston, McGill-Queen's University Press.
McNeill, David. 2005. *Gesture and Thought*, Chicago, The University of Chicago Press.
Marraud, Hubert. 2018. *Arguments from Ostension*, Argumentation 32, 309–327.
Misak, Cheryl J. 1995. *Verificationism. Its History and Prospects*, London and New York, Routledge.
Nance, Dale A. 2016. *The Burden of Proof: Discriminatory Power, Weight of Evidence, and Tenacity of Belief*, Cambridge, Cambridge University Press.
Osborne, Martin J. & Ariel Rubinstein. 1994. *A Course in Game Theory*, Cambridge (Mass.), The MIT Press.
Peirce, Charles S. 1998. *The Essential Peirce. Vol. 2 (1893–1913)*, ed. by the Peirce Edition Project, Bloomington and Indianapolis, Indiana University Press.
Putnam, Hilary. 1975. *Mind, Language and Reality. Philosophical Papers, vol. 2*, Cambridge, Cambridge University Press.
Quine, Willard V. O. 1960. *Word and Object*, Cambridge (Mass.), The MIT Press.
Short, Thomas L. 2004. *The Development of Peirce's Theory of Signs*. In Cheryl Misak (ed.), *The Cambridge Companion to Peirce*, Cambridge, Cambridge University Press, 214–240.
Short, Thomas L. 2007. *Peirce's Theory of Signs*, Cambridge, Cambridge University Press.
Sperber, Dan & Deirdre Wilson. 1995. *Relevance. Communication and Cognition*, 2$^{nd}$ ed., Oxford, Blackwell.
Tiercelin, Claudine. 2011. *Le Ciment des choses. Petit traité de métaphysique scientifique réaliste*, Paris, Ithaque.
Tiercelin, Claudine. 2018. *Et si nous considérions l'abduction comme un sentiment épistémique?*, Recherches sur la philosophie et le langage 34, 87–102.
Tiersma, Peter M. & Lawrence M. Solan (eds.). 2012. *The Oxford Handbook of Language and Law*, Oxford, Oxford University Press.
Tuzet, Giovanni. 2021a. *The Pragmatics of Evidence Discourse*. In Christian Dahlman, Alex Stein & Giovanni Tuzet (eds.), *Philosophical Foundations of Evidence Law*, Oxford, Oxford University Press, 169–182.

Tuzet, Giovanni. 2021b. *Testimony and Hearsay*. In Verena Klappstein & Maciej Dybowski (eds.), *Theory of Legal Evidence – Evidence in Legal Theory*, Cham, Springer, 205–223.

Tuzet, Giovanni. 2022. *On Probatory Ostension and Inference*. In Jordi Ferrer & Carmen Vázquez (eds.), *Evidential Legal Reasoning: Crossing Civil Law and Common Law Traditions*, Cambridge, Cambridge University Press, 138–158.

Tuzet, Giovanni. 2023. *Expert Signs and Legal Burdens*, International Journal for the Semiotics of Law 36, 159–183.

Twining, William. 1984. *Evidence and Legal Theory*, Modern Law Review 47, 261–283.

Walton, Douglas, Fabrizio Macagno & Giovanni Sartor. 2020. *Statutory Interpretation. Pragmatics and Argumentation*, Cambridge, Cambridge University Press.

Wittgenstein, Ludwig. 1953. *Philosophical Investigations*, ed. by Gertrude E. M. Anscombe & Rush Rhees, Oxford, Blackwell.

Section 2: **Looking at language to investigate legal challenges**

Daniel Greineder
# Illusions of a common Language: Impressions of an arbitration practitioner

**Abstract:** International arbitration resolves a wide range of mostly commercial disputes globally. Parties freely choose a procedural language. This chapter gives a non-technical introduction and describes some of the linguistic, legal and cultural challenges facing users, who often adopt English.

**Keywords:** International arbitration, legal English, forced monolingualism, pragmatics

## 1 Introduction

This paper sets out some impressions of the challenges of language use facing arbitration practitioners. Its author is mindful of having been very much a non-specialist at the hospitable 5th General Convention of the International Language and Law Association in 2021 in Alicante and as much a potential object of academic study as a contributor. Nonetheless, it is hoped that the thoughts sketched here will act as a fillip to practitioners to reflect on the use of language in arbitration and an invitation to linguists to explore some of its peculiarities.

It may seem odd to charge lawyers with linguistic insensitivity, when their work is inescapably verbal. The lawyer of the popular imagination illustrates this, whether it is the ponderous practitioner caught up in cant and prolixity, or the silver-tongued advocate whose turn of phrase wins hopeless cases. However, neither shows any capacity for reflecting on language. The former wallows in it, while the latter rarely rises above shallow virtuosity. Consider, too, that lawyers are victims of declining standards of the teaching of grammar in any language. It is not uncommon for a lawyer to argue the correct construction of a contract term without being able to parse it. Nor are such solecisms as "the passive tense" unknown even in the London High Court.

Lawyers are then as susceptible as others to the phenomenon identified in linguistics that the user is so exposed to language and uses it so easily as to be blind to its peculiarities. "Phenomena can be so familiar that we really do not see them at all." (Chomsky, $^3$2006, 21) To an extent the mysteries of language hide in

**Daniel Greineder,** Geneva

https://doi.org/10.1515/9783110799651-005

plain sight from its users. International arbitration is a special case of legal language use. It is, as the name suggests, highly international in bringing together parties, witnesses, counsel, arbitrators and expert witnesses from different countries, often in the same proceedings. It faces the many challenges that arise when the law must speak a foreign language. For example, it is necessary to take witness evidence, where the witness is not a native speaker of the procedural language and may speak that language poorly or depend entirely on an interpreter. Not only do such difficulties present themselves in sometimes extreme form in arbitration, but there is the added peculiarity that there is no official language of arbitration as there is, for example, of proceedings before the German or English courts. In principle, parties are free to choose the procedural language and may choose a language different from the language of the law governing dispute. It is not unusual for parties in dispute over a contract governed by Swiss law to choose English as the language of the proceedings. This means that legal arguments will be made and witness evidence taken in English. The award, setting out the arbitral tribunal's final analysis and decision, will be in English, too.

## 2 What is international arbitration?

There is no shortage of legal textbooks explaining international arbitration for those seeking a detailed introduction. The leading commentaries include: Blackaby/Partasides/Redfern, [7]2022; Born, [3]2020; Gaillard, 2010. For immediate purposes, a much-simplified account should suffice. At its simplest, arbitration is based on the idea that two parties to a dispute may choose a judge – or arbitrator – to decide their dispute and will also agree to accept the judge's binding decision. They do so by means of a contract, the arbitration agreement, sometimes referred to imprecisely as the "arbitration clause". This voluntary agreement leads to a binding outcome. In contrast to proceedings before state courts, no one is forced to arbitrate and the arbitrator's authority, unlike the judge's, derives entirely from the will of the parties.

In commercial disputes, especially with an international element, arbitration has become the leading means of dispute resolution as a form of privatized litigation, where international parties choose to opt out of the jurisdiction of the state courts in favour of a procedure that is supposedly more suited to their needs and is neutral. The national arbitration laws of many jurisdictions allow parties to do this and give them the freedom to choose not only their arbitrators but also many of the procedural rules, subject to minimal safeguards of fairness and equal treatment.[1] This means that the parties and, without their agreement, the arbitrator may decide whether to hold a hearing, call witnesses, invite written submissions, or require the production, disclosure or discovery of documents – irrespective of the approach of the local domestic courts.

In international commerce, the attractions are clear. Arbitration is also much used to resolve disputes in other sectors, notably sports law and investor-state claims brought under treaties. Such cases may be brought under the auspices of ICSID, the Permanent Court of Arbitration or CAS. The procedure enables parties to create a bespoke tribunal for the unique purpose of deciding their dispute. As Webster (2009), among others, explains, a tribunal will come into being to decide a particular dispute referred to it and then cease to exist, becoming *functus officio* after it has done so. If the same parties run into a further dispute, they will typically start from scratch and appoint a new arbitral tribunal.

Arbitration also offers the prospect of neutrality. Where, for example, a Korean and a U.S. company entered into a joint venture to carry out a construction project in Nigeria, neither would be happy to refer disputes to the courts of the other for fear of being at a tactical and cultural, if not legal, disadvantage. Nor would either party necessarily want to submit disputes to the Nigerian courts, where proceedings have been known to be rather protracted. Moreover, the contract establishing the joint venture may be governed by a neutral law, such as, in this case, English law.

An arbitral tribunal with its seat, for example, in Geneva, i.e., the place whose local courts have oversight over the arbitration, will better serve their needs. Each party will be free to participate in the appointment of an arbitral tribunal. In a typical larger arbitration, for example, where the amount in dispute exceeds USD 5 million, there will be three arbitrators, one appointed by each party and a chairman or presiding arbitrator, jointly appointed by the parties or by a neutral third party, such as an arbitration institution. Some of the mechanisms are set out in the following rules: ICC Rules Article 12, UNCITRAL Arbitration Rules, Articles 8 to 10, and LCIA Arbitration Rules, Article 5 to 8. The parties may each choose someone of their own nationality, or with experience in the particular business sector, or sympathetic to their business culture. The point at which a judge who is sympathetic to a party's point of view, for example, by virtue of sharing a certain culture, disqualifies himself for being biased or lacking

---

[1] For illustrations of these minimal but vital safeguards, see Arbitration Act 1996, Section 34; Code de procédure civile, Article 1511; Loi Fédérale Suisse sur le droit international privé, Chapitre 12 – Arbitrage International, Article 182. 2017 ICC Rules of Arbitration, Article 19; 2014 LCIA Rules, Article 14.5; 2012 Swiss Rules of International Arbitration, Article 15; 2018 DIS Arbitration Rules, Article 21.3; SIAC Rules 2016, Article 27.

independence, is a delicate question. For the most part, arbitration lives comfortably with the tension between a chosen arbitrator's openness to a party's peculiar position and that arbitrator's independence and capacity for dispassionate judgement. Whether this is justified or merely convenient may be open to debate. Any international award, the arbitral tribunal's decision at the outcome of an arbitration, will be enforceable internationally in accordance with the Convention on the Recognition and Enforcement of Foreign Arbitral Awards, known as "the New York Convention," to which almost all states globally are signatories. By this means, states recognize foreign awards and enable parties to enforce them.

The continuing success of arbitration as the preferred means of resolving disputes depends on many factors. Some of these are legal, for example, the ease of recognition and enforcement of awards in various jurisdictions, i.e., how easy is it is for the winning party obtain money from the losing party; others depend on practical matters, notably the cost and efficiency of proceedings. Of greater interest to students of linguistics is perhaps the question whether the global appeal of arbitration should rest on a single set of best practices for all users, or whether arbitration should encourage diverse practices internationally. To put it in linguistic terms, should arbitration offer a legal Esperanto or an all-inclusive polyglot polyphony?

The debate arises in relation to procedural practices. Generally, parties to arbitration have considerable freedom in organizing the proceedings as they see fit. For example, they may want proceedings that resemble court proceedings in their domestic courts. Alternatively, they may have chosen arbitration for the very reason that they did not feel well-served by domestic court practices. They wanted something different.

The leading arbitration laws and institutional rules offer parties that choice.[2] The French practitioner Mathieu de Boisséson (2014, 520) describes that freedom as a "vide fructueux" or fertile void. It has been filled by a variety of institutional rules as well as good practice guidelines drafted by committees of arbitral luminaries and promoted by professional associations, sometimes known as "soft law" instruments. As Greineder/Medvedskaya (2020), Kaufmann-Kohler (2010), Park (2006), and Schneider (2011) show in different ways, such regulation is controversial and often eludes conceptualization. Principally, these are the IBA Guidelines on Conflicts of Interest, the IBA Rules on the Taking of Evidence, IBA Guidelines

---

[2] By way of illustration, the reader is referred to the following : Arbitration Act 1996, Section 34; Code de procédure civile, Article 1511; Loi Fédérale Suisse sur le droit international privé, Chapitre 12 – Arbitrage International, Article 182. 2017 ICC Rules of Arbitration, Article 19; 2014 LCIA Rules, Article 14.5; 2012 Swiss Rules of International Arbitration, Article 15; 2018 DIS Arbitration Rules, Article 21.3; SIAC Rules 2016, Article 27.

on Party Representation, the ILA Recommendations on Res Judicata and Arbitration, and the recent Prague Rules. Importantly, these rules do not have the authority of legislation or even contractual obligations that parties may voluntarily assume. Rather, they are recommendations by panels of more or less expert experts. The battle for procedural supremacy is controversial. Are the guidelines on procedural practices actually any good? Do they favour one paradigm over another, such as common law practices over civil law practices? What authority and legitimacy do the rules and those who promulgate them have? The argument has a largely overlooked linguistic aspect: who determines the procedural terminology of arbitration and with what result? Can and indeed should there be a neutral international drafting style? Unlike in the case of domestic legislation, there is no established legal language or terminology. The institutions and committees that draft soft law must create their own terminology. Anecdotally, such rules are never fun to read and on occasion troublingly opaque. They also raise underexplored linguistic questions, notably whether the preponderance of English in international arbitration entails the imposition of an English law culture (see Greineder, 2021).

## 3 What is the language of arbitration?

If arbitration does not have an official language, what then is the language of arbitration? The simplest answer is that the language of an arbitration is the language chosen by the parties. Most arbitration clauses, such as those published by the arbitral institutions, contain a provision stipulating the language or occasionally languages of the proceedings. The choice is often determined by practical considerations. These may include the party representatives' native languages, their likely choice of counsel – especially, whether they will be working with typically anglophone international law firms –, the availability of suitably qualified arbitrators, fluent in a particular language, as well as of expert witnesses, such as accountants who help to formulate damages claims, and the convenience of taking witness evidence in one language or another. Sometimes tactical considerations will play into this. For example, one party may press to conduct proceedings in a neutral language, such as English, that is the native language of neither party.

On a deeper level, the answer is more difficult. Esperanto might have commended itself as a single official language, but it never caught on. The resultant vacuum is filled by whatever happens to be the dominant language. The statistics of some of the leading institutions, which administer and provide some organizational framework for arbitrations, suggest that English is a popular choice. For example, SCC statistics for 2021 record that well over 40% of proceedings were conducted in English, while the ICC statistics for 2019 record that 79% of awards were rendered in English, indicating that English was the language of the proceedings. This is unsurprising given its dominance as a language of global commerce. Parties will often choose to resolve their disputes in the language in which they originally did business.

The dominance of English is a snapshot of current global society and business. It is unclear from the statistics, yet quite possible, that the choice of English is especially prevalent among high-value claims over USD 100 million, because largely anglophone law firms dominate those cases. This would give it extra significance as a language of arbitration. 40 years ago, French would have played a greater role, although perhaps more for legal than economic reasons. French jurists made a vital contribution to modern theories of arbitration. In 20 years' time, Spanish and Mandarin may have risen to greater prominence. The dominance of English may be accounted for in four ways. English has come to be a lingua franca of world commerce. Further, unlike in the case of Latin in the early modern period, there are many native speakers of the language. Many of them are economically highly active globally. The common law with its origins in England spread to many jurisdictions through the British Empire, including large parts of Asia and Africa. Most law in the U.S. was originally part of that heritage, and New York law remains a popular choice for international contracts. Finally, leading law firms whose main offices are in London or the U.S. have branched out offering global legal services and exporting legal English on the back of English as a language of commerce. According to Global Arbitration Review in 2021, the closely followed annual GAR 30 ranking of leading arbitration practices listed only two law firms based outside the English-speaking world, one of which, the Swiss firm Lalive, has an office in London.

How any of the leading languages has fared in international arbitration may be a worthwhile object of linguistic study. The importance of Spanish, for example, owes more to the many disputes originating in South America rather than to any connection to Spain. Some leading U.S. law firms, such as King & Spalding, now advertise the Spanish-speaking capability of their arbitration lawyers. In the German-speaking world, there seems to be an extraordinary zeal for conducting proceedings in English, although the language of the applicable laws – German, Swiss or Austrian – is German and German legal terminology with its origins in civil law culture does not translate readily into English.

## 4 English and the price of dominance

This paper is limited to some reflections on English as a language of arbitration. These are, as the title indicated, impressions of a practitioner who avails himself of the amateur's privilege to venture into conjecture. It is suggested that English as a widespread, but by no means universal, language of arbitration has contributed to the development of international arbitral practices and strengthened it institutionally. A more or less common and widely spoken language facilitates communication. However, users' levels of proficiency in English and the development of English as a language of arbitration have not kept apace with the internationalization of arbitration itself. The result is a form of legal English that is often unsubtle and deficient, and thus a victim of the success of arbitration. The use of English in arbitration eludes empirical study because arbitrations are largely confidential. It is not usually possible for third parties to gain access to the case file.

Two trends run parallel in the use of English in arbitration, harmonization and fragmentation; of these, harmonization predominates. In the absence of an official language and established terminology, arbitration draws freely on different sources to establish a common idiom. These include that of domestic legal systems. For example, since cross-examination in arbitration and court practice do not differ fundamentally, there is no need to coin a new term. Predictably, authors of key textbooks, law firms and arbitrators are freer to make their mark on legal usage. In particular, the leading international law firms, such as those listed in the GAR 30, have developed a distinctive style that is prevalent in high-value disputes. It reappears to a degree in awards, for which arbitrators draw on written submissions. Regrettably, as most proceedings are confidential, it is difficult to provide examples.

Surprisingly, law firm arbitration English is not especially technical in its legal usage and sometimes almost colloquial. It avails itself liberally of lively, if sometimes strained similes, and appeals to fairness and good faith. Baseball references creep in quite often to the predictable incomprehension of at least some of the readership. This informality may be to accommodate the difficulty that the lawyers writing the submissions have not always studied the applicable law of the dispute. They are not fully competent to deploy its terminology. For example, it is not unusual to instruct a leading New York or London firm in a dispute arising in a CIS country and governed by the law of that country. International counsel, perhaps admitted to the bar in England or New York, will be assisted by local lawyers but will in practice control much of the drafting. Moreover, the arbitrators themselves may have different legal backgrounds, one may be Swiss, another French and the third English. A fact-based argument, highlighting the equities of the case, may provide a better common denominator from counsel's point of view than legal arguments that each arbitrator would approach differently by virtue of a different legal training. The arbitrators may feel more confident basing their decisions on factual arguments than legal niceties.

The style often has a mid-Atlantic ring. Parties may file briefs (American) rather than submissions (English) but make applications (English) rather than bring motions (American). The results are often bland because they are multi-authored and heavily edited for stylistic consistency. The idiomatic style is easy to digest but difficult for non-native speakers to adopt. Especially when writing informally, native speakers assert their dominance over the language of the proceedings. It may be of interest to students of arbitration and linguistics that apparent accessibility and informality are elitist in excluding many users from the dominant idiom. Mid-Atlantic drafting breeds a peculiar pastiche when, as often happens, French or German practitioners try to adopt it, banging figuratively on the table to complain of "bad faith" and failures of "fair dealing" as if they were in a New York court.

Arbitration English may be pedestrian but is not easy to master. Not everyone can keep up. In that case, arbitration practice adopts several responses. Witness evidence is assimilated to arbitration English even where the witness is incapable of speaking it. Much witness evidence is provided to arbitral tribunals in the form not of oral evidence at the hearing but of written witness statements, later tested in oral cross-examination. These are prepared on the basis of witness interviews by lawyers acting for one of the parties. Quite often, even native speakers could not have written the lawyerly statements as they are submitted. The texts are more heavily edited by lawyers than the witness and usually shoehorned into a house style of the relevant law firm. To give the text an authentic ring, the witness may be permitted one or other unidiomatic phrase. This supposedly reassures the arbitrators that the authorial voice is that of a foreigner. The approach does not, however, protect a witness from embarrassment at the hearing. It is not unusual for witnesses to struggle to explain their own written evidence in oral testimony.

Oral witness evidence poses problems of its own. The arbitration community has yet to come to terms with the difficulties of oral witness evidence given through an interpreter. There is a pool of well-qualified interpreters who are used to arbitration hearings, but even so some languages are far better served than others. It would compromise access to justice if, in English language arbitration, a witness whose evidence was in Russian, were better served by the interpreters than a Mandarin-speaking witness in the same proceedings who also testified through an interpreter. Yet, anecdotally, this seems to be the case. Nor are arbitrators always sufficiently sensitive to cultural factors affecting witness testimony. Interpreting is not simply a matter of matching words but also of cultural contexts. Generally, a Japanese salaryman speaking through an interpreter in the presence of his manager at the hearing will give evidence differently from an American working in sales and used to giving flamboyant pitches. The tendency of arbitrators is to form a superficial impression of whether the interpreters were "good" and then to read the hearing transcript as if it were the original statement of the witness.

Hearing transcripts are a common feature of hearings. Modern court reporting services, such as Opus 2, can provide the participants of a hearing with a fairly polished transcript within a couple of hours of the end of the hearing day. They also use software, such as LiveNote, which enables participants to follow the hearing on screen. This is an essential aid to many arbitrators and counsel whose first language is not English and who may struggle with a witness whose accent and locution are unfamiliar. While this software is undoubtedly useful, for example, when challenging a witness on a point in his oral testimony, heavy reliance on it leaves both oral evidence and submissions, such as opening and closing statements, impoverished. Non-verbal communication is devalued in favour of textualized advocacy. The effect is reminiscent of following a film by reading the subtitles at the expense of the spoken word and images.

To accommodate these difficulties users of arbitration may simplify their usage or limit their range of reference. On occasion, international arbitration cruelly exposes the limitations of global culture – at least, as it is known to lawyers. In one case, lead counsel, not necessarily an *homme de lettres*, bewildered his opponents by using the term "Kafkaesque". He did so in its vulgar sense to express horror at bureaucratic complexity and opacity rather than to denote things pertaining to the life and work of Franz Kafka. There was a visible rush to Wikipedia in the rows of his opponents from a leading U.S. law firm. On other occasions, in the quest for the lowest common denominator, practitioners may stoop very low indeed. In one case, a counsel team rejected the catchphrase "Documents are like diamonds: they are for ever," but only after much self-congratulation on the literary and cinematic allusion to a well-known British secret agent.

Where it becomes impossible to sustain the pretence of a common language, English as a language of arbitration fragments into Englishes, as practitioners adopt different forms of the language. The terminology of English shipping and reinsurance arbitrations in the London market is deeply rooted in English legal culture. Few foreigners enter the market. Equally, Swiss arbitration English has followed a distinct path. Generally, the English of the Zurich arbitration scene, with its slightly academic Germanic ring, is of a good standard but also distinctive in having evolved with only limited involvement of native speakers. Nor have the anglophone international firms made their mark in Switzerland, where they are barely present, as they have in Paris and Frankfurt. Swiss lawyers draft at one remove from the jargon of their anglophone counterparts. Such separation engenders fragmentation. An English barrister would struggle with the linguistic transition from the High Court in London to an Ur-Swiss M&A dispute in provincial Bern.

Occasionally, English language arbitration falls back on a second language. This is to be distinguished from the case of bilingual arbitration, where parties expressly adopt more than one procedural language. Where, in an English-language arbitration, a contract is governed by German law and the language of the proceedings is English, it is common to appoint German-speaking and German-qualified arbitrators.[3] The translation of German legal terminology is difficult. Rather, than rely on awkward translations, the arbitrators may ask for legal exhibits to be submitted in the original and fall back on the original language of the law. Deliberations among the arbitrators will also take place in German. The phenomenon of an unofficial second language in arbitration is common in European civil law jurisdictions, where commercial considerations favour English but legal considerations favour the use of the authentic language of the law.

The examples discussed illustrate that great difficulties of communication lie behind the statistics about the popular choice of English as a procedural language of arbitration. Sometimes the choice is wildly aspirational, as counsel are blatantly unfit to make their case in English; on other occasions, it is only practicable because users of arbitration cast a blind eye on serious difficulties of communication. The linguistic stakes of arbitration look set to rise. Institutions are keen to follow the practice of ICSID, which administers many investor-state arbitrations, in publishing awards that would otherwise remain confidential. The reasons for this are not always clear. It may be a way of making arbitration more accountable to the world at large, a means of encouraging a doctrine of precedent, or a marketing tool for institutions to showcase the breadth of their caseload. In any case, arbitral awards that previously only determined disputes between parties will become public and some form of reference for other tribunals. It remains to be seen whether the language of arbitration is sufficiently precise for arbitral awards to be used in those ways by third parties, who are unfamiliar with the case and seeking generally applicable guidance.

It is no criticism of practitioners that they must deal with languages they speak poorly or not at all. Even though the number who can conduct an arbitration in two languages is significant and the number who can do so in three small but not negligible, many practitioners are inevitably confronted with languages

---

[3] For an erudite and practical discussion of the difficulties of the related question of drafting contracts in English from a German lawyer's perspective, see Triebel/Vogenauer, 2018.

largely or entirely foreign to them. To some degree, some complications, distortions and misunderstandings are inescapable. It is, however, culpable to make light of the compromises and shortcomings in the use of language so as to maintain the fiction of international arbitration as having some semi-official and universal language or group of languages. The political philosopher Isaiah Berlin liked to paraphrase Kant that out of the crooked timber of humanity nothing entirely straight could be made. It is the lawyer's lifelong task to accept the crookedness of the available materials and make the most of them. To the linguist considering the use of language in arbitration it may be an invitation and a challenge.

# Bibliography

## Books and Articles

Blackaby Nigel, Constantine Partasides & Alan Redfern. [7]2022. *Redfern and Hunter on International Arbitration*, Oxford, Oxford University Press.

Born, Gary B. [3]2020. *International Commercial Arbitration*, Alphen aan den Rijn, Kluwer.

Chomsky, Noam. [3]2006. *Language and Mind*, Cambridge, Cambridge University Press.

De Boisséson, Mathieu. 2014. *La "Soft law" dans l'arbitrage*, Cahiers de l'arbitrage, 519–524.

Gaillard, Emmanuel. 2010. *Legal Theory of Arbitration*, Leiden, Nijhoff.

Greineder, Daniel & Anastasia Medvedskaya. 2020. *Beyond High Hopes and Dark Fears: Towards a Deflationary View of Soft Law in International Arbitration*, ASA Bulletin 38, 414–435.

Greineder, Daniel. 2021. *Wirrsal, Esperanto und Kulturkampf: zum Englischen als Rechts- und Verfahrenssprache in der internationalen Schiedsgerichtsbarkeit*. In Daniel Greineder, Karl Pörnbacher & Stefan Vogenauer (eds.), *Schiedsgerichtsbarkeit und Rechtssprache: Festschrift für Volker Triebel zum 80. Geburtstag*, München, C.H. Beck, 43–61.

Kaufmann-Kohler, Gabrielle. 2010. *Soft Law in International Arbitration: Codification and Normativity*, Journal of International Dispute Settlement 1, 283–299.

Schneider, Michael. 2011. *The Essential Guidelines for the Preparation of Guidelines, Directives, Notes, Protocols and Other Methods Intended to Help International Arbitration Practitioners to Avoid the Need for Independent Thinking and to Promote the Transformation of Errors into "Best Practices"*. In Yves Derain & Laurent Lévy (eds.), *Liber Amicorum en l'honneur de Serge Lazareff*, Paris, Pedone, 563–567.

Park, William W. 2006. *The Procedural Soft Law of International Arbitration: Non-Governmental Instruments*. In Loukas A. Mistelis & Julian D. M. Lew (eds.), *Pervasive Problems in International Arbitration*, Alphen aan den Rijn, Kluwer, 141–154.

Triebel, Volker & Stefan Vogenauer. [1]2018. *Englisch als Vertragssprache: Fallstricke und Fehlerquellen*, München, C. H. Beck.

Webster, Thomas H. 2009. *Functus Officio and Remand in International Arbitration*, ASA Bulletin 27, 441–465.

*The GAR 30 and Power Index Revealed*, Global Arbitration Review, 01.07.2021 (globalarbitrationreview.com/article/the-gar-30-and-power-index-revealed) (11.02.2022).
Global Arbitration News (https://globalarbitrationnews.com/wp-content/uploads/2020/07/ICC-DR-2019-statistics.pdf) (11.02.2022).
SCC Statistics (https://sccarbitrationinstitute.se/en/about-scc/scc-statistics-2021) (21.02.203).

## Laws and Statutes

Treaties
Convention on the Recognition and Enforcement of Foreign Arbitral Awards (1958) (also known as "the New York Convention")
England and Wales
Arbitration Act 1996
France
Code de procédure civile
Switzerland
Loi Fédérale Suisse sur le droit international privé

## Rules of Arbitral Institutions and Soft Law Instruments

2018 DIS Arbitration Rules
IBA Guidelines on Conflicts of Interest
IBA Rules on the Taking of Evidence
IBA Guidelines on Party Representation
2021 ICC Rules of Arbitration
ILA Recommendations on Res Judicata and Arbitration
LCIA Arbitration Rules (effective 1 October 2020)
Rules on the Efficient Conduct of Proceedings in International Arbitration (Prague Rules)
SIAC Rules 2016
2021 Swiss Rules
UNCITRAL Arbitration Rules (2013 version)

Jacqueline Visconti
# Pragmatic features of Italian court proceedings

**Abstract:** Clarity of court proceedings is an essential tool for both the efficiency and the quality of a modern judiciary. However, the language in court is often obscure. This issue can probably be referred to in any legal system (cf. e.g. Kimble 2006), but it is particularly problematic in Italy due to a long-standing tradition that encourages unduly verbose texts (cf. e.g. Mortara Garavelli 2001). Aiming to create a new resource for clear and effective writing of court proceedings, the *ClearAct* PRIN project was funded in 2018 by the *Ministero dell'Università e della Ricerca*. The purpose of the present paper is to report on the first results of this project. Firstly, the semi-automatic annotation method for anonymising personal data specifically devised for the project will be described. Secondly, the composite nature and pragmatic properties of these documents will be highlighted: Who are the recipients? How are the many voices rendered that echo in these documents?

Avenues for further research are indicated, such as an enlargement of the scope of the investigation to other judicial systems and international courts, as clarity and concision have bearings also in the realm of international judicial cooperation.

**Keywords:** Clarity of court proceedings, language in court, language of the defence counsel, legal language

# 1 Introduction

Clarity of court proceedings is increasingly seen as an essential tool for both the efficiency and the quality of a modern judiciary. In Italy, for example, countering a long-standing tradition that encourages a convoluted style of writing (cf. e.g. Mortara Garavelli 2001), a series of measures, including statutory, case- and soft law, have set clarity and concision of proceedings among the fundamental principles of the trial.[1] Thus, in the recent law of 26 November 2021, n. 206 (article 17.d),

---

[1] For an overview cf. Canzio (2022, 277–279); Gualdo (2021, 13); Pandimiglio-Masia (2020, 145–148).

**Jacqueline Visconti,** Università di Genova

https://doi.org/10.1515/9783110799651-006

compliance with the principles of clarity and concision is explicitly referred to in the provisions aimed at faster and more efficient civil cases.

As is known, the drive towards clarity and concision is fostered by international impulses: stringent indications are given by the European Court of Human Rights and by the Court of Justice of the European Union both to national courts and to the parties of the case: «Owing to the need to translate it into all the official languages of the European Union, the request for a preliminary ruling should therefore be drafted simply, clearly and precisely, avoiding superfluous detail».[2] Similarly: «In the interests of the proper conduct of the procedure and in the interests of the parties themselves, the written pleadings or observations must therefore be drafted in clear, concise language, without the use of technical terms specific to a national legal system. Repetition must be avoided and short sentences must, as far as possible, be used in preference to long and complex sentences that include parenthetical and subordinate clauses».[3]

Clear/plain language has been the subject of extensive investigation both in common and civil law traditions.[4] Common law experts have elaborated on key aspects of legal writing: word choice, punctuation, grammar and syntax, rhetorical figures.[5] However, the debate on plain language has mainly focused on legislative drafting, while judicial writing has been concerned only to a lesser extent.[6] Here, the various judicial traditions – in line with diverse cultures – favour different styles of exposition. As for the writings of judges, for example, the individualised, more immediate and colloquial style of the common law world (in some way marked by orality, sometimes with striking expressions) differs from the continental style (French, German, Italian), which is decidedly more solemn, formal, and impersonal.[7]

A significant experience in this field was promoted by the Italian Ministry of Justice, with the establishment in 2016 of a working group on the concision of court proceedings, coordinated by the then Head of the department for Justice affairs Antonio Mura. The group, in accordance with the multi-disciplinary nature of the topic, included jurists as well as linguists. The overarching purpose was to

---

[2] *Recommendations to national courts and tribunals in relation to the initiation of preliminary ruling proceedings* (2012/C 338/01, 21).
[3] *Practice directions to parties concerning cases brought before the Court*, Official Journal of the European Union (LI/42, 14.02.2020, 42).
[4] Cf. Williams (2018) and references therein for an updated discussion.
[5] Cf. e.g. Ferreri (2019); Garner (2002); Williams (2022).
[6] Cf. Kimble (2006) or Duckworth (1994) on "plain judicial language", and more recently, Williams (2020) on plain language developments in Canada and UK court judgments.
[7] A deeper scrutiny would show significant differences also within each of these traditions in their actual developments in different countries.

improve the efficiency of the Italian judiciary through more concise and clearer court proceedings, thus allowing citizens to better understand the functioning of the legal system, and thereby enhancing the democracy of justice.[8]

## 2 The *ClearAct* project

A particularly under-researched area is the language of counsel proceedings, i.e. texts written by defence advocates. Compared to several important studies on judgments, only a few have focused on defence documents.[9]

The reasons for this are mainly practical and methodological: while there are numerous databases of judgments, a similar archive for counsel proceedings does not yet exist. The *dossier* is normally inaccessible to the non-lawyer scholar (the few studies mentioned used personal collaborations with law firms).

Yet, such texts play a fundamental role in the economy of the trial: since each phase is influenced by the preceding ones and reverberates on the following ones in an organic sequence, it is fundamental that rigour and sobriety be there right from the introductory proceedings.

It is no coincidence that the Italian *Ministero dell'Università e della Ricerca* recently funded a national PRIN project on *La chiarezza degli atti del processo (AttiChiari)*: (Clarity in Court Proceeding – ClearAct) an unpublished database for the scholar and the citizen.[10]

The project aims at creating a new resource for an effective writing of court proceedings written by the defence, by building a synchronic corpus of about three million words and a searchable database. The specific objectives are: (i) the collection, for the first time in Italy, of a database of such counsel documents, relating to proceedings both by the Italian Supreme Court (*Corte di Cassazione*) and by a set of local Courts and Courts of Appeal; (ii) the qualitative study of the pragmatic features and the rhetorical and argumentative structure of these texts.

The documents collected amount so far to around two million words, concerning civil and, to a lesser extent, criminal and administrative matters.

---

[8] Cf. the report: www.federnotizie.it/wp-content/uploads/2018/10/CHIAREZZA_ATTI_PROCES SUALI.pdf, as well as Mura/Visconti (2022); Visconti (2022, 26ff.).
[9] In particular, for Italy: Sabatini (2003); Mortara Garavelli (2003); Gualdo/Dell'Anna (2016); Dell'Anna (2016).
[10] Cf. the project website (https://attichiari.unige.it) and Gualdo/Clemenzi (2021); Dell'Anna (in press) for an outline of the first results of the research.

## 2.1 An-tool

The main difficulty in collecting such documents is that they contain personal data, the disclosure of which would violate the right to confidentiality of the parties involved in the proceedings. Since the documents are provided by counsels on a voluntary basis, a tool had to be devised that removes anthroponyms, toponyms, dates and any other personal data.

Anonymisation practices are often used to erase personal data before judgments are made available to the public, especially when dealing with sensitive subjects (e.g. involving minors) or when sensitive data occur in the text.[11]

Traditional anonymisation practices consist in simply removing data by erasing them, by replacing them with asterisks, *omissis*, letters, or other graphic signs, or by leaving blank spaces, as shown below (Candrilli 2021, 21):

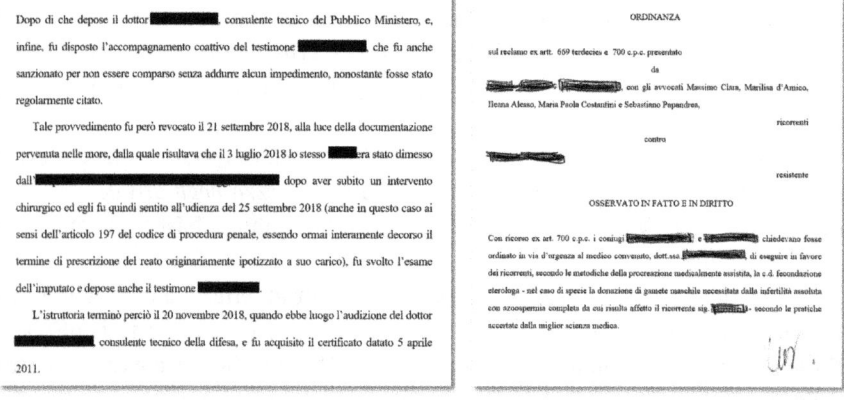

**Fig. 1:** Traditional anonymisation practices.

These practices, however, clearly affect the readability of the documents, thus hindering the possibility of linguistic analyses, such as the reconstruction of anaphoric chains, the study of variation or of other textual phenomena. Readable and complete texts are yet essential for both linguistic and legal studies, which aim to investigate the pragmatic and textual characteristics of the documents, e.g. the strategies used by counsels in referring to the assisted party, the counterpart and third parties involved in the trial.

---

[11] The terms "personal data" and "sensitive data" are defined in the Reg. EU 2016/679, artt. 4 and 9.

Thus, for the purposes of the project, a new method for anonymising personal data has been devised, which consists in a semi-automatic annotation method for the pseudonymisation of court documents (Fusi 2021).

Pseudonymisation is defined by the General Data Protection Regulation (Reg. EU 2016/679) as «the processing of personal data in such a manner that the personal data can no longer be attributed to a specific data subject without the use of additional information, provided that such additional information is kept separately and is subject to technical and organisational measures to ensure that the personal data are not attributed to an identified or identifiable natural person».

As illustrated in detail in Clementi *et al.* (in press), the research team developed a software that automatically replaces personal data with fictional data of the same category extracted from predefined lists, after applying a light markup to the texts. This software also makes it possible to maintain replacements consistent within the same text, or within several texts relating to the same judgement.

The annotation process consists of two stages: manual annotation, followed by automatic replacement of the personal data (Fusi 2021). As specified by Clementi *et al.* (in press), firstly, the operator manually annotates the source text directly in a word-processing application by applying a specifically devised syntax, which indicates both the category of the sensitive data and the genre; the portion of text to be replaced is inserted in curly brackets, preceded by a tag indicating its category (anthroponym, toponym, date, etc.) on a {tag:text} pattern, as in the example shown below:[12]

---

[12] Tags include: a-f-f (*anthroponym, female, first*) for female anthroponyms; a-m-f (*anthroponym, male, first*) for male anthroponyms; j-f (*juridical person, female*) for legal entities of feminine grammatical gender; j-m (*juridical person, male*) for legal entities of masculine grammatical gender; t (*toponym*) for toponyms, etc. (for the complete list, cf. Fusi 2021).

**TRIBUNALE DI {t:ROVIGO}**

**R.G. N. {n:1234}/{d:2015}**

**PER**

La {j-f:Prima} Spa, in persona del legale rappresentante {f-lat:pro tempore}, Sig. {a-m-f:Mario} {a-l:Rossi}, con sede in {t:Rovigo}, Via {ad:Giuseppe Garibaldi n. 23}, P.IVA {n:01234567890}, rappresentata e difesa dall'Avv. {a-f-f:Ada} {a-l:Verdi} (C.F. {u:VRDDAA67A41H620P}, fax {n:0425/123456}, pec {m:adaverdi@pec.it}) presso il cui Studio sito in {t:Rovigo}, Via {ad:Guglielmo Marconi n. 55}, ha eletto domicilio giusta delega posta in calce al presente atto

**CONTRO**

Il Sig. {a-l:Grossi} {a-m-f:Zeno} con l'Avv. {a-f-f:Alma} {a-l:Piccoli}

**COMPARSA DI COSTITUZIONE E RISPOSTA**

Con atto di citazione notificato in data {d:10/10/2014}, il Sig. {a-l:Grossi} {a-m-f:Zeno} conveniva in giudizio dinanzi l'intestato Tribunale la {j-f:Prima} Spa per sentire accertare e dichiarare la responsabilità della società convenuta per i vizi di costruzione insistenti nel garage n. {n:99} sito in {t:Rovigo} di proprietà dell'attore. Per l'effetto, il Sig. {a-l:Grossi} chiedeva che il Tribunale adito condannasse la {j-f:Prima} Spa a risarcire tutti i danni, patrimoniali e non, subiti dallo stesso attore e, in particolare, a titolo di danno patrimoniale, alla eliminazione dei vizi costruttivi da individuarsi e quantificarsi con apposita consulenza tecnica, alla somma di € {n:900,00} quale differenza tra i danni materiali risarciti e quelli richiesti in sede di mediazione e alla somma di € {n:3.000,00} a titolo di ulteriori danni patrimoniali ovvero da liquidarsi in via equitativa; a titolo di danno non patrimoniale il Sig. {a-l:Grossi} chiedeva che la {j-f:Prima} Spa venisse condannata alla somma di € {n:3.500,00} ovvero a quella diversa di liquidarsi in via equitativa, e dunque complessivamente alla somma di € {n:7.400,00} oltre alla eliminazione dei vizi costruttivi, ovvero alla somma che sarà ritenuta di giustizia.

**Fig. 2:** Annotated text.

For each category of personal data identified by a tag, the programme draws on a list of terms of the same category (male and female first names, surnames and toponyms) to automatically replace them (the programme chooses names beginning with the same initial as the original ones to preserve euphony). Therefore, marking *Prima* as "juridical person" (j-f) instructs the programme to remove this item and replace it with another legal person name from the list provided.

The result of the pseudonymisation process, as can be seen below, is a coherent text that can be fruitfully analysed from a textual perspective:

TRIBUNALE DI ROGOLO

R.G. N. 6183/2002

PER

La Perla Spa, in persona del legale rappresentante pro tempore, Sig. Maccabeo Renzullo, con sede in Rogolo, Via Jazelynn Tundis, 88, P.IVA 35068033487, rappresentata e difesa dall'Avv. Algeri Valeri (C.F. PTKLTZ49P50U275B, fax 4841/007983, pec ip6042@gmail.com) presso il cui Studio sito in Rogolo, Via Yilin Uggetti, 80, ha eletto domicilio giusta delega posta in calce al presente atto

CONTRO

Il Sig. Gardumo Zosimo con l'Avv. Adelaide Piccioni

COMPARSA DI COSTITUZIONE E RISPOSTA

Con atto di citazione notificato in data 15/10/2001, il Sig. Gardumo Zosimo conveniva in giudizio dinanzi l'intestato Tribunale la Perla Spa per sentire accertare e dichiarare la responsabilità della società convenuta per i vizi di costruzione insistenti nel garage n. 84 sito in Rogolo di proprietà dell'attore. Per l'effetto, il Sig. Gardumo chiedeva che il Tribunale adito condannasse la Perla Spa a risarcire tutti i danni, patrimoniali e non, subiti dallo stesso attore e, in particolare, a titolo di danno patrimoniale, alla eliminazione dei vizi costruttivi da individuarsi e quantificarsi con apposita consulenza tecnica, alla somma di € 387,67 quale differenza tra i danni materiali risarciti e quelli richiesti in sede di mediazione e alla somma di € 4.273,81 a titolo di ulteriori danni patrimoniali ovvero da liquidarsi in via equitativa; a titolo di danno non patrimoniale il Sig. Gardumo chiedeva che la Perla Spa venisse condannata alla somma di € 4.749,39 ovvero a quella diversa di liquidarsi in via equitativa, e dunque complessivamente alla somma di € 9.023,21 oltre alla eliminazione dei vizi costruttivi, ovvero alla somma che sarà ritenuta di giustizia.

**Fig. 3:** An-tool pseudonymised text.

# 3 Pragmatic features of Italian counsel proceedings

Italian court proceedings consist of a variety of documents, from those initiating the trial (*atto di citazione, comparsa di risposta*) to the judgment (*sentenza*). Most cases continue in an appeal (*appello*), followed by a higher court judgment, and into a further appeal to the *Corte di cassazione* (*ricorso*) (last year appeals to this Supreme Court amounted to over 70.000 cases!). For this reason, Italian court proceedings are characterized by a "vertical" intertextuality (Dell'Anna/Gualdo 2016), i.e. by a relationship not only with the relevant legislation and case law, but with a whole series of preceding documents that are relevant to the case, such as the introductory proceedings, the first degree judgment, *etc.*, in a chain of texts mirroring one into the other.

To add to their textual complexity, most of these documents have a composite nature, comprising at least four textual types: narrative and explanatory, in the reconstruction of the facts of the case and of the previous phases of the trial; argumentative and (in judgments only) prescriptive. Their structure reflects the complexity of legal reasoning, as manifested in chains of subordinate propositions; the lexicon comprises technical terms, but also obsolete words and Latinisms.

A further feature shaping the language in Italian court proceedings is that of having institutionally more than one addressee: both defence counsels and the judge know that their words, although directed to citizens, will be subject to the careful reading of their colleagues; as a consequence, as noted, among others, by Bellucci (2005, 453) and Gualdo (2018, 232), court proceedings in Italy are not written to be understood by the layman, but by a pool of experts, be it the counsel for the counterpart or the judge.

When comparing texts drafted by the judge with texts drafted by defence counsels, a few differences appear to be salient: firstly, as they aim to convince the judge, the latter are characterised by a greater personalisation and by the pervasive presence of connotative elements; exclamatory and interrogative propositions are often inserted, as an expression of a "dialogic mode" (Gualdo/Dell'Anna 2016, 630) that reproduces in writing the interaction with the judge within the hearing. Secondly, while defence documents end with a request to the judge (*Voglia l'illustrissimo tribunale . . .* 'May the illustrious court be willing to . . .'), judgments obviously end with the decision.

In the remainder of this section, I shall focus on defence proceedings. Although the investigation is only at the beginning, as the elaboration of both the *corpus* and the search engine is still in progress, the pragmatic texture of court proceedings appears to be shaped by the following three phenomena.

The first feature concerns the polyphonic nature of these texts. As is known, court proceedings echo a whole choir of voices: the parties in the case, the preceding judge, the legislator, the experts, in a patchwork of recordings of both monologues and dialogic exchanges, which sometimes took place orally and were only later verbalized in writing.[13] Correspondingly, a whole battery of linguistic and textual devices is employed to render such a polyphonic texture: direct and indirect speech, of course, but also reportive conditional mood, as in: *secondo la ricostruzione avversaria, il Codice della Strada non sarebbe applicabile* ('according to the counterpart, the *Codice della strada* would not be applicable'), adjectives and adverbs marking what is said as someone else's voice, such as *asserito* ('alleged'), *preteso* ('purported'), *presunto* ('presumed'), or punctuation marks having the

---

13 Cf. e.g. Gualdo 2018, 233. In general, on the "dialogicity" of written texts, cf. Calaresu (2022).

function of reporting the counterpart's words, as in: *completa ed ingiustificata mancata concessione della giornata festiva per 'shopping' o 'influenza'* (complete and unjustified failure to meet parental requirements due to 'shopping' or 'flu').[14]

A second relevant feature concerns the degree in which the speaker/writer's presence is manifested in these texts, which, as noted by Turco (in press), may surprise in an exposition of a legal-deductive nature.[15] Indicators of the speaker/writer's presence are frequent in counsel proceedings, as these texts have also a persuasive function: epistemic and evaluative adverbs (*sicuramente* 'undoubtedly', *purtroppo* 'unfortunately', *fortunatamente* 'luckily'), intensifiers (*fortemente* 'strongly'; *copioso contratto* 'copious contract', *interminabili fasi successive* 'endless subsequent phases'), parenthetical comments, such as *Il tempo (mesi!) passava . . .* 'Time (months!) went by', rhetorical figures, such as *climax: Il credito vantato dalla ricorrente è senza dubbio certo ed esigibile, nonché supportato da idonea prova scritta ed anzi incontestato* 'The credit claimed by the plaintiff is undoubtedly certain and collectible, as well as supported by appropriate written evidence and indeed undisputed'.

When completed, the *ClearAct* database will allow a quantitative assessment of this phenomenon, also in relation to the hypothesis of a gradient, ranging from the language of the counsel to the judgment. In particular, following the approach outlined in Ricci/Rossari (2016), the presence of the speaker/writer can be quantified by measuring the systematic co-occurrence of devices expressing it, for instance evaluative adverbs, argumentative connectives, *etc.* A statistic study of the combinatory modalities of these devices would allow estimating the degree of locutor's presence in the text (degree of "subjectivity"): the more an item reflects the voice of the speaker/writer, the more it will co-occur with other forms displaying the same feature.

Finally, a third property of these texts has not, to my knowledge, been brought to attention: the role of implicit linguistic devices.

Recent research focusing on the persuasive and manipulative power of linguistic implicit communication highlights the reduction in critical vigilance that such implicit communication provokes (cf. e.g. Lombardi Vallauri/Cominetti/Masia 2022 and references therein): in other words, when content is conveyed implicitly, it is less likely to be challenged or even questioned by addressees. One can see how the language of the defence counsel may avail itself of such powerful persuasive techniques.

---

[14] On the function of punctuation marks in Italian counsel proceedings cf. Lombardi (in press).
[15] On the notion of "subjectivity" and on the fundamental role of modulation as "part and parcel" of utterance production and reception cf. e.g. Caffi (2005, 13ff.).

A clear instance of this phenomenon are presuppositions triggered by additive focus adverbs, such as *anche* 'too',[16] or by constructions such as *oltre a* 'as well as', *ulteriore* 'further'. As is well known, additive elements convey the implicit content that the asserted proposition holds for other entities too. When a defence advocate writes that 'damages were recorded also as a consequence of a later subsequent sewage suction operation, with further contamination of movable and immovable property',[17] presupposed content, such as the recorded damages and the soiling of the client's goods, are presented as taken for granted. Similarly, when she/he writes that 'in addition to the water and sanitary system, the event caused problems in the electrical system, especially at the ground sockets',[18] presupposed content, such as 'the event caused damages to the sanitary and water system', will be taken for granted, whereas asserted content, such as 'the event caused problems to the electric system', is more likely to be subject to rebuttal.

Through implicit devices such as these, the defence counsel manages to insert in the text, and in the case, a set of granted elements, thus subtracting them to potential rebuttal.

# 4 Conclusions

The obscurity of legal language has been subject to criticism in almost every tradition.[19] As already spelled out by Jonathan Swift:

«This Society hath a peculiar Cant and Jargon of their own, that no other Mortal can understand, and wherein all their Laws are written [. . .] whereby the have wholly confounded the very Essence of Truth and Falsehood, of Right and Wrong . . .» (*Gulliver's Travels* 1726).

Yet in the Anglo-saxon world too, as said above, court language has been touched only marginally by the drive towards transparency brought by *plain language movement*, which has mainly focused on the language of the law.

In a recent book, Williams (2020) discusses a few examples of judgments written in *plain language*. In such cases, relating to English and Canadian courts, the

---

**16** Cf. De Cesare (2022) on such presuppositions in traditional and social media.
**17** *Si sono registrati danni anche in conseguenza del successivo intervento di aspirazione dei liquami, con ulteriori imbrattamenti ai beni mobili ed immobili* (*ClearAct*, civ-ge-tro-702bis-201911_01).
**18** *Oltre che all'impianto idrico-sanitario l'evento ha cagionato problematiche all'impianto elettrico, soprattutto in corrispondenza delle prese a terra* (*Clear Act*, civ-ge-tro-702bis-201911_01).
**19** Cf. e.g. Visconti (2018, 1).

language reflects the wish of the judge to be understood by the addresses, who are children, in some cases, Canadian indigenous people in others.

Although so far only qualitative, the analysis of the pragmatic features of counsel court proceedings offered in this chapter may contribute to understanding the complexity of such texts, with the aim of reducing unnecessary intricacies. To this purpose, the *ClearAct* database will hopefully pave the way for the elaboration of similar resources in other languages and legal traditions.[20]

# Bibliography

Bellucci, Patrizia. 2005. *La redazione delle sentenze: Una responsabilità linguistica elevata*, Diritto e Formazione, 5/3, 447–465.

Caffi, Claudia. 2005. *Mitigation*, Leiden, Brill.

Calaresu, Emilia. 2022. *La dialogicità nei testi scritti*, Pisa, Pacini.

Candrilli, Fernanda. 2021. *Il progetto di archiviazione e anonimizzazione*. In Riccardo Gualdo & Laura Clemenzi (eds.), *Atti Chiari. Chiarezza e concisione nella scrittura forense*, Viterbo, Sette Città, 19–29.

Canzio, Giovanni. 2022. *Dire il diritto nel XXI secolo*, Milano, Giuffrè.

De Cesare, Anna Maria. 2022. *Opinion shaping in the context of the "Me Too" movement. An investigation of presuppositions triggered by additive focus adverbs in traditional and social media*, Journal of Pragmatics 188, 1–13.

Dell'Anna, Maria Vittoria. 2016. *Fra attori e convenuti. Lingua dell'avvocato e lingua del giudice nel processo civile*. In Federigo Bambi (ed.), *Lingua e processo. Le parole del diritto di fronte al giudice*, Firenze, Accademia della Crusca, 83–101.

Dell'Anna, Maria Vittoria (ed.). (in press). *Lingua e scrittura forense. Storia, temi, prospettive*, Torino, Giappichelli.

Duckworth, Mark. 1994. *Clarity and the rule of law: the role of plain judicial language*, Judicial Review: Selected Conference Papers, Journal of the Judicial Commission of New South Wales, Vol. 2, No. 1, 69–88.

Ferreri, Silvia. 2019. *Falsi amici nelle corti. Leggere le sentenze di Common Law evitando le trappole linguistiche*, Torino, Giappichelli.

Fusi, Daniele. 2021. *Digitalizzazione e marcatura XML degli atti*. In Riccardo Gualdo & Laura Clemenzi (eds.), *Atti Chiari. Chiarezza e concisione nella scrittura forense*, Viterbo, Sette Città, 59–73.

Garner, Bryan A. 2002. *The Elements of Legal Style*, Oxford, Oxford University Press.

Gualdo, Riccardo. 2018. *Elogio della raccomandazione. Analisi linguistica di sentenze e relazioni della Corte dei conti*, Studi linguistici italiani, XLV, 231–273.

Gualdo, Riccardo. 2021. *Chiarezza e concisione negli atti processuali*, in Riccardo Gualdo & Laura Clemenzi (eds.), *Atti Chiari. Chiarezza e concisione nella scrittura forense*. Viterbo, Sette Città, 11–18.

---

**20** Cf. Visconti (2023) for a proposal of a comparative investigation on UK and Italian court documents.

Gualdo, Riccardo & Laura Clemenzi (eds.). 2021. *Atti Chiari. Chiarezza e concisione nella scrittura forense*, Viterbo, Sette Città.
Gualdo, Riccardo &Maria Vittoria Dell'Anna. 2016. *Per prove e per indizi (testuali). La prosa forense dell'avvocato e il linguaggio giuridico*. In Giovanni Ruffino & Marina Castiglione (eds.), *La lingua variabile nei testi letterari, artistici e funzionali contemporanei. Analisi, interpretazione, traduzione*, Firenze, Cesati, 623–635.
Kimble, Joseph. 2006. *Lifting the Fog of Legalese. Essays on Plain Language*, Durham NC, Carolina Academic Press.
Lombardi, Giulia. (in press). *Verso la redazione di Attichiari: Il ruolo della punteggiatura*. In Maria Vittoria Dell'Anna (ed.), *Lingua e scrittura forense. Storia, temi, prospettive*, Torino, Giappichelli.
Lombardi Vallauri, Edoardo, Federica Cominetti & Viviana Masia. 2022. *The persuasive and manipulative power of implicit communication*, Journal of Pragmatics 197, 1–7.
Mortara Garavelli, Bice. 2001. *Le parole e la giustizia. Divagazioni grammaticali e retoriche su testi giuridici italiani*, Torino, Einaudi.
Mortara Garavelli, Bice. 2003. *Strutture testuali e stereotipi nel linguaggio forense*. In Alarico Mariani Marini (ed.), *La lingua, la legge, la professione forense*, Milano, Giuffrè, 3–19.
Mura, Antonio &Jacqueline Visconti. 2022. *Concision and Clarity in Court Proceedings*. In Gianluca Pontrandolfo & Stanislaw Gozdz (eds.), *Law, Language and the Courtroom. Legal Linguistics and the Discourse of Judges*, London, Routledge, 231–242.
Pandimiglio, Matteo Viviana Masia. 2020. *Linguistica, diritto e variazione: uno sguardo al linguaggio delle sentenze in Italia*. In Jacqueline Visconti, Manuela Manfredini & Lorenzo Coveri (eds.), *Linguaggi settoriali e specialistici*, Firenze, Cesati, 145–148.
Ricci, Claudia & Corinne Rossari. (2023). *L'apporto dell'analisi quantitativa nello studio linguistico di alcune forme dell'argomentazione*. In Angela Ferrari, Letizia Lala & Filippo Pecorari (eds.), *Forme della scrittura italiana contemporaneo in prospettiva contrastiva. La componente testuale*, Firenze, Cesati.
Sabatini, Francesco. 2003. *Dalla lingua comune al linguaggio del legislatore e dell'avvocato*. In Alarico Mariani Marini & Maurizio Paganelli (eds.), *L'avvocato e il processo. Le tecniche della difesa*, Milano, Giuffrè, 3–14.
Turco, Simone. (in press). *Oggettività, soggettività, narratività. Per un approccio narratologico agli atti di parte*. In Maria Vittoria Dell'Anna (ed.), *Lingua e scrittura forense. Storia, temi, prospettive*, Torino, Giappichelli.
Visconti, Jacqueline (ed.) in collaboration with Monika Rathert. 2018. *Handbook of Communication in the Legal Sphere*, Berlin, De Gruyter Mouton.
Visconti, Jacqueline. 2022. *Studi su testi giuridici: norme, sentenze, traduzione*, Firenze, Accademia della Crusca.
Visconti, Jacqueline. (2023). *Aspetti della testualità in atti giudiziari italiani e inglesi*, In Angela Ferrari, Letizia Lala & Filippo Pecorari (eds.), *Forme della scrittura italiana contemporaneo in prospettiva contrastiva. La componente testuale*, Firenze, Cesati.
Williams, Christopher. 2018. *Legal drafting*. In Jacqueline Visconti (ed.), *Handbook of Communication in the Legal Sphere*, Berlin, De Gruyter Mouton, 13–35.
Williams, Christopher. 2022. *The impact of plain language on legal English in the United Kingdom*, London, Routledge.
Williams, Christopher. 2020. *Court judgments in plain language. Some recent developments in Canada and United Kingdom*. In Stefania Maci, Michele Sala & Cinzia Spinzi (eds.), *Communicating English in specialised domains: A Festschrift for Maurizio Gotti*, Newcastle, Cambridge Scholars, 106–125.

Jakub Eryk Marszalenko
# Politeness Matters: What honorifics can tell us about accuracy in Japanese-English court interpreting

**Abstract:** One of the aspects of the Japanese language that makes it different from English in a considerable way is the robust and complex honorific system in the former language, referred to as *keigo*. This system manifests itself in different layers of the Japanese language, such as the lexis, inflection, and pragmatics. It is particularly prominent in verbs, which in spoken Japanese take either casual or formal inflection forms. Moreover, this can be further complicated by the fact that many verbs and verb forms have inherent honorific value, either humble or respectful, which is selected depending on the communication act and its interactants. Such various forms are not easily translatable into English and pose a challenge in fields of translation in which accuracy is as sacrosanct an aspect as court interpreting. This article addresses how court interpreters in Japan working with English deal with *keigo*, what choices they make, and how these choices may impact our understanding of accuracy and equivalence.

**Keywords:** court interpreting, honorifics, *keigo*, Japanese-English translation, accuracy

## 1 Introduction

Honorifics, or more broadly, politeness, are an integral part of a speaker's linguistic repertoire regardless of their language or the sociocultural environment in which this language is used. However, the application of honorific strategies varies across languages and cultures, requiring participants in interlingual and intercultural settings to be aware of such differences. Furthermore, even within the same speech community, "honorifics can be manipulated, and may dynamically change in accordance with given contexts and the interactants' stances, en-

---

**Note:** This research project was supported by: JSPS KAKENHI Grant Number: 20K13038 and The Promotion and Mutual Aid Corporation for Private Schools of Japan (PMAC): Scholarship for Young Researchers.

---

**Jakub Eryk Marszalenko,** Nagoya University of Foreign Studies

https://doi.org/10.1515/9783110799651-007

hancing the quality of communication" (Obana 2020, 248). This implies that such linguistic phenomena are not static or fixed but depend entirely on the circumstances surrounding the communication act and its participants.

The Japanese language has a highly complex honorific system, which differs from that of English considerably. According to Obana (2020, 256), unlike the honorific strategies in English used mainly to mitigate 'face-threatening acts,' such strategies employed in Japanese are predominantly concerned with one's *tachiba* in the communication act. *Tachiba* refers to any sociocultural circumstances determining the interactants' position in the act, such as social hierarchy and status, the relationship between the interlocutors, and the level of formality of the conversation.

Consequently, translating between Japanese and English (in both directions) can pose considerable challenges, especially in fields and settings where precision and accuracy are paramount. Court interpreting is a prime example of such a setting, giving the court interpreter little (if any) "poetic license" in their rendition of the target text. As the *Hōtei Tsūyaku Handobukku* (*Court Interpreting Handbook*), published under the supervision of the Supreme Court of Japan, informs prospective and practicing court interpreters, due to the importance of witness testimony given in court, "(. . .) it is necessary to interpret all utterances *word for word*" (Hōsōkai 2011, 1; emphasis added).[1]

However, it must be acknowledged that "[i]nterpreters are required to make instantaneous and irrevocable decisions about how to interpret from the source language. Inevitably, the interpreter will sometimes choose an inappropriate equivalent term, or there will be ambiguities in the source language which allow more than one valid interpretation (Laster/Taylor 1994, 183)." Furthermore, depending on the language pair involved in the translational act, such choices by court interpreters may also be conditioned by the fact that translating the source text "word for word" into the target language is either simply not possible or undesirable, because it would hinder, rather than enhance, the accuracy of translation and, consequently, communication between the speakers of the two languages.

This article deals with situations where the court interpreter faces a choice derived precisely from such differences. This challenge will be discussed with Japanese as both the source- and the target language. Examples of renditions by court interpreters from actual interpreter-mediated criminal trials where English was used will be provided to address this issue (several additional monolingual cases will also be introduced in Section 2). The trial observations took place between June 2020

---

[1] Unless specified otherwise, all translations from Japanese and Polish into English were made by the author.

and June 2022 and included twenty criminal cases heard in the following district courts throughout Japan: Tokyo (seven cases), Naha (in Japan's southernmost prefecture of Okinawa; seven cases), Osaka (three cases), Chiba (two cases), and Yokohama (one case). The nationalities of the non-Japanese-speaking defendants and witnesses included the United States (twelve persons), Nigeria (six persons), and Canada (two persons). One of the reasons for such a large proportion of American nationals in the dataset is that seven of the twenty observed trials took place at Naha District Court on the island of Okinawa, which is home to a considerable number of US military personnel and their contractors, as well as their families.

The examples of translation provided in the subsequent parts of this article will demonstrate that different honorific strategies employed by English and Japanese speakers make the demand for a "word-for-word" rendition futile because "pragmatic factors [such as politeness] are overtly encoded for to a greater level in Korean and Japanese than in English (Kiaer/Cagan 2023, xvi)." As a consequence, when Japanese is the source language, certain intricacies and nuances of honorifics are lost or sacrificed in the English rendition due to the constraints put on court interpreters. Kiaer and Cagan refer to this phenomenon as "pragmatic invisibility" and note that "[p]ragmatic "invisibles" are most prevalent in translation between languages that do not share some cultural consensus (Kiaer/Cagan 2023, 7)." In contrast, with English as the source language, the court interpreter needs to decide which Japanese honorific forms to choose, possibly impacting the listeners' (the Japanese-speaking participants of the trial: judges, lay judges, prosecutors, and defense lawyers) perceptions of the source-text speaker (the defendant or a witness speaking English).

The examples will also highlight a conflict that may arise in court interpreting: the discrepancy between expectations exemplified by the instructions in the *Court Interpreting Handbook* (Hōsōkai 2011) quoted earlier on the one hand, and the complex reality of the translational act on the other. As a court interpreter interviewed for the author's previous study notes, "(. . .) when legal practitioners talk of 'faithful interpretation,' they take this mathematical standpoint that there is only 'one correct translation.' From a practicing interpreter's perspective, though, it seems that there is a certain range for 'correct translation' (Marszalenko 2016, 37)." Indeed, the examples presented in the subsequent parts of this article will be, in the author's view, within this "range of correct translation," especially because, "[i]deally, [legal] interpreting should be user-centric and take into account text, subject, culture, time, purpose, and context-dependent aspects (Kadrić 2021, 507)." Japanese honorifics are a prime example of a situation wherein culture, purpose, and context, among other factors, play a role of a degree hard to overestimate. Therefore, the fact that such renditions may seem controversial (i.e., that interpreters took "excessive liberty" in producing them), if, indeed, they do seem so, could serve as an argument that the understanding of the complexities of court interpreters' work is still insufficient.

## 2 Honorific strategies in Japanese

### 2.1 *Keigo:* The Japanese honorific language

One of the most significant differences between the Japanese honorific system and those in most European languages (including English) is that no sentence in (spoken) Japanese can be neutral in terms of its honorific value (Huszcza 2006, 155). This is because, as Kiaer and Cagan note, "[i]t would not be going too far to say that one cannot confidently form an utterance in Korean or Japanese without an idea of who is speaking to whom (Kiaer/Cagan 2023, xv)." For example, Huszcza (2006, 155) gives the example of a sentence in Polish (*Dzisiaj jest sobota*), which can be easily translated into English as 'Today is Saturday.' This sentence is honorifically neutral in both languages and does not need to change depending on who its addressee is. However, the same sentence can be translated into Japanese in various ways, depending on who is uttering it to whom. More specifically, the relationship between the speaker and the addressee and the conversational situation determine the level of politeness and formality that needs to be applied in conveying a piece of information as seemingly simple and neutral as the day of the week.

The Japanese honorific expressions are usually referred to collectively as *keigo*. This linguistic system can be broadly divided into three main categories: *sonkeigo, kenjōgo,* and *teineigo. Sonkeigo* (lit. 'respectful speech') is used to elevate the status of the hearer or those in their circle, such as their family, friends, and persons in their workplace. In contrast, *kenjōgo* ('humble speech') is used when the speaker wishes to lower their own status or the status of those in their circle. *Teineigo* can be used by speakers of (mostly) equal status in somewhat formal situations but not where being overly respectful or humble is necessary.

This honorific feature of the Japanese language manifests itself in many aspects of speech, such as the lexicon and inflection (of verbs, in particular). In the lexical realm, for example, the word for 'family' in Japanese can be rendered neutrally (or, depending on the situation, casually) as *kazoku* or honorifically as *go-kazoku*, with *go-* as an honorific prefix added to nouns when the family being referred to is worthy of respect owing to, for example, being the family of the hearer of the utterance.

A similar phenomenon can be observed with other parts of speech, such as verbs, where one can encounter lexemes with different honorific values. Thus, a sentence such as 'Have you eaten yet?' will be rendered differently depending on who the addressee of the question is:

Example 1
English: Have you eaten yet?

Example 1.1
Japanese (casual): *Mō tabeta (ka)?*

Example 1.2
Japanese (formal+respectful): *Mō meshiagarimashita (ka)?*

In the two Japanese sentences above (1.1 and 1.2), different verbs are used for the English 'to eat.' The verb used in Example 1.1 (*tabeta*) is the past tense of the less formal *taberu* in its casual form (its formal inflection style in the present tense is *tabemasu* and in the past tense *tabemashita*). In contrast, the verb in Example 1.2 (*meshiagarimashita*) is the past tense of the honorific *meshiagaru* in its formal inflection. These formal inflection styles, which Obana (2020) refers to as 'addressee honorifics,' are commonly known as the *desu/masu*-style named after the inflection forms of verbs (i.e., the *-masu* in *tabemasu* or *meshiagarimasu*). The *ka* in the examples is a question particle, which in spoken Japanese is sometimes omitted.

Thus, we can surmise that the speaker in Example 1.1 is talking to a friend or a family member in an informal situation (replacing *taberu* with *kuu* would make the utterance more casual still or even vulgar and give it a distinctly masculine character). In contrast, the speaker in Example 1.2 is likely addressing someone of a social status superior to their own, such as a teacher or supervisor, or trying to convey politeness and respect to their interlocutor in a more formal setting.

The Japanese honorific system is too robust and complex to provide a comprehensive description here. Still, several more examples will help understand court interpreters' challenges when rendering Japanese honorifics. To simplify this description, only spoken Japanese will be discussed. Furthermore, in the subsequent parts of this article, the discussion on Japanese honorifics in interpreter-mediated criminal trials will be limited to the use of verbs.

In the discussion about the honorific value of verbs, we need to distinguish between a verb's inherent honorific value and the honorific value of the verb's inflection form. In addition to honorifically neutral verbs such as *taberu*, there are also those which are considered humble: HUM (*itadaku*: 'to eat,' 'to drink' or 'to receive'), or respectful: RES (*meshiagaru*: 'to eat' or 'to drink') in themselves. Furthermore, certain inflection forms of verbs can also manifest an honorific value, either informal/casual: CAS or formal/polite: POL. Thus, in spoken Japanese, a verb can be respectful in itself but used in a casual (informal) inflection style, e.g., *Sensei ga meshiagatta* (RES-CAS) translating to 'The professor ate' (RES-CAS). The same sentence in the formal (polite) inflection style takes on the form *Sensei ga meshiagarimashita* (RES-POL) without changing the utterance's meaning.

This honorific value of the inflection form is the reason for the lack of honorifically neutral utterances in spoken Japanese alluded to previously. More specifically, any utterance containing a verb must either be informal/casual (CAS) or formal/polite (POL). It must be noted, however, that a verb usually takes the formal inflection style only when it is the main verb (predicate) of the sentence. Verbs in subordinate clauses may be used in a form similar to the English infinitive, which is often identical to the informal inflection style.

In Table 1 below, the verb *kaku* ('to write') illustrates the multiple honorific strategies available to speakers. It must be borne in mind that one could point to various exceptions in the usage of the forms presented here. Moreover, the humble and respectful forms can be used toward a direct addressee (the hearer of the utterance) or those in their circle. However, for clarity, the table was compiled with the premise that the 'protagonist of the utterance' (Huszcza 2006), in this case, the agent performing the action of writing, is either the speaker or the addressee.

**Table 1:** Forms of the verb *kaku* ('to write') on different levels of formality and with varying honorific values.

| Informal Inflection (CAS) | Formal Inflection (POL) | | | Protagonist |
|---|---|---|---|---|
| | Honorific Value: Polite (POL) | Honorific Value: Humble (HUM-POL) | Honorific Value: Respectful (RES-POL) | |
| *kaku* (CAS) | *kakimasu* (POL) | – | – | Speaker or Addressee |
| *o-kaki-suru* (HUM-CAS) | – | *o-kaki-shimasu* (HUM-POL) | – | Speaker |
| *kakasete itadaku* (HUM-CAS) | – | *kakasete itadakimasu* (HUM-POL) | – | Speaker |
| *Kakareru* (RES-CAS) | – | – | *Kakaremasu* (RES-POL) | Addressee |
| *o-kaki-ni naru* (RES-CAS) | – | – | *o-kaki-ni narimasu* (RES-POL) | Addressee |
| *o-kaki-ni narareru* (RES-CAS) | – | – | *o-kaki-ni nararemasu* (RES-POL) | Addressee |

The informal inflection forms can be used in speech in less formal (casual) situations (i.e., among friends, close colleagues, and family members).² The formal inflection forms are used when some degree of formality and politeness is expected or required.

In the complex honorific system presented above, the same form of a verb (for example, the respectful *o-kaki-ni naru* [RES]) can be used in situations requiring different levels of inflectional formality. For example, when speaking directly to 'Professor Tanaka,' a student may ask, *Kono hon wa, Tanaka-sensei ga <u>o-kaki-ni narimashita</u> (ka)?* ('Did you write [RES-POL] this book, Professor Tanaka?'). In contrast, the student may ask their classmate the same question using a more informal inflection form (CAS) but retaining the level of respect toward the protagonist of the utterance, i.e., Professor Tanaka: *Kono hon wa, Tanaka-sensei ga <u>o-kaki-ni natta</u> (ka)?* ('Did professor Tanaka write [RES-CAS] this book?'). In this latter example, the speaker is talking casually (CAS) to the addressee (their classmate) but is using the respectful verb form (RES) referring to the protagonist of the utterance (Professor Tanaka).

In summary, the Japanese language provides multiple options for its speakers regarding honorific strategies. Furthermore, such strategies can be applied at various layers of the utterance. The honorific style of Japanese conversations is not fixed or rigid. Speakers shift from one honorific strategy to another depending on the circumstances, such as who their interlocutor is, how well they know them or how close they are, and even how friendly the atmosphere of the conversation is. In addition, gender, age, or social status are also factors that can play a role in determining the appropriate honorific style. Moreover, the same speaker may apply different honorific strategies toward the same interlocutor in the same communication act, depending on what effect they wish to achieve. What matters for this discussion is, to repeat Huszcza's (2006, 155) comment, that in spoken Japanese, no sentence is neutral in terms of its honorific value, and the seemingly neutral utterances (i.e., informal inflection forms in Table 1) should be treated as casual and inappropriate in more formal situations.

## 2.2 Honorific strategies in the courtroom

Before diving into honorific strategies applied by court interpreters in criminal trials where English is used, we first need to investigate what strategies can be found in monolingual trials where all participants speak Japanese. This will be

---

2 In written Japanese, the same forms are often treated as 'plain' or honorifically neutral. However, such usage will not be addressed in this article.

done based on several examples of interactions observed in criminal cases in different courtrooms throughout Japan (these cases are separate from the interpreter-mediated trials enlisted in the *Introduction*).

In the actual courtroom discourse, the formal style (formal inflection forms in Table 1) is the default form of predicates. This implies that a certain level of formality is retained due to the nature of the proceedings, but interactants treat each other as equals or at least wish to convey the atmosphere of equality or fairness or preserve the discourse decorum fit for the courtroom setting by using this linguistic style. For example, during the opening procedures of a criminal trial, the presiding judge asks the defendant the following question to verify their identity: *Namae wa nan to iimasu ka* ('What is your name?'), where *iimasu* is the formal inflection style (POL) of the verb *iu* ('to say' or to 'call').

However, the "default" honorific style does not imply that the interactants in a trial must abide by it in all situations. Further, the notion of equality in terms of one's *tachiba* in a trial does not necessarily reflect the facts on the ground. To begin with, the legal practitioners in the courtroom are clearly more knowledgeable and better equipped to deal with the legal matters at hand and, thus, have 'epistemic primacy' over the lay participants of the trial (see Shimotani 2012 and Yonezawa 2021 for a discussion on how this may impact the speech style in Japanese courtrooms). In contrast, the defendant or witnesses usually are not legal experts, and their knowledge of the law and legal strategies is arguably limited. More importantly, defendants in a criminal trial are in a far more disadvantaged position because their freedom or life is at stake, and the lawyers (especially judges and prosecutors) wield substantial power over their fate.

### 2.2.1 Examples of usage of informal inflection (CAS)

Given the above imbalance of power among the participants in criminal trials, one could expect that if a deviation from the default formal inflection occurs, it will cause the speakers with the higher *tachiba* (i.e., legal practitioners) to speak more casually to defendants or witnesses, and the latter interactants to speak more formally when addressing such "high-ranked" hearers. Employing such honorific strategies can indeed be observed in criminal trials, as in the examples below, in which the defense counsel (DC) addresses his client, the defendant.

Example 2
DC: *Hanzai no kikkake wa, dō iu koto datta* (CAS) *ka ne?*
(Translation: 'I wonder, what was [CAS] the reason that led to the crime?')

Example 3
DC: *Hontōni hansei shitoru* (CAS) *yo ne?*
(Translation: 'You are truly remorseful [CAS], right?')

The two examples above come from the same case observed at Nagoya District Court in central Japan. The underlined portions (*datta* in Example 2 and *hansei shitoru* in Example 3) are casual inflection forms (CAS) of verbs. *Datta* is the past tense of the verb *dearu* (in certain situations similar to the English 'to be'), whose more formal inflection (POL) style is *desu* in the present tense or *deshita* in the past tense. Similarly, *hansei shitoru* in Example 3 is a casual and abbreviated form of the verb *hansei suru* ('to feel remorse,' 'to be remorseful') in its continuous form, which turns into *hansei shite iru* in the unabbreviated version. In the formal style, this verb form is replaced with *hansei shite imasu* rather than *-iru*.

It would have also been appropriate for the defense counsel to use a more formal style in his questions. Why, then, did this lawyer feel comfortable using these casual forms? 'Epistemic primacy' mentioned above may be one reason, but it does not explain this choice entirely. Another possible factor in the style choice is the considerable age difference between the defense counsel and his client. The defendant was in his early twenties, whereas the defense counsel appeared to be in his sixties. The age difference, however, can only play such a role if the speaker does indeed possess epistemic primacy over the hearer. In other words, if the situation were reversed (i.e., if the defense counsel was much younger than the defendant), it would be unlikely for the defendant to use the casual style when addressing their defense counsel and, consequently, the Court.

## 2.2.2 Examples of usage of formal inflection (POL)

The opposite phenomenon, namely, using a more formal linguistic style, can also be observed in Japan's monolingual courtroom discourse.

As mentioned earlier, gender and age can play a role in determining the level of formality and honorific style of the utterance. In general, a more formal and polite style is associated more with women than men. However, this description is a simplification and may not reflect the natural speech patterns of all speakers (see Inoue 2003 for a detailed discussion on the issues associated with Japanese "women's language") and may lead to gender-based stereotyping. Furthermore, it is impossible to determine whether the speaker's gender informs a particular speech style in a given conversational act or if the style is idiosyncratic to that speaker, with their gender being only a minor factor.

With the above limitations to treating gender as a factor determining one's speech style in mind, let us turn to several examples from a fraud case against two male defendants (72 and 57 years old, respectively) heard at Naha District Court (Okinawa), in which the presiding judge (J) was a woman in her middle age. In the following examples, the judge addresses both defendants simultaneously.

Example 4
J: *Kisojō wo uketotte oraremasu* (RES-POL) *ka?*
(Translation: 'Have [you] received [RES-POL] the indictment act?')

Example 5
J: *Mazu, watashi no hō kara shitsumon sasete itadakimasu* (HUM-POL).
(Translation: 'First, let me ask [HUM-POL] [you] some questions.')

In both instances, the judge uses the more formal, honorific speech style. Not only do these forms belong to the formal inflection (POL) shown in Table 1 (this, by itself, would not make the style humble or respectful, but merely "polite"), but also *sonkeigo* (RES, Example 4) and *kenjōgo* (HUM, Example 5) forms of verbs.

*Uketotte oraremasu* in Example 4, uses the continuous honorific form *orareru* in its formal inflection style. Two strategies were applied here: first, the judge replaced the more neutral continuous verb form *iru* with *orareru* (RES), and second, she used it in its formal inflection style (POL). The same sentence in a less honorific but still sufficiently formal style could be *Kisojō wo uketotte imasu* (POL) *ka?*

In Example 5, the judge uses a humble but, at the same time, typical style of speech in everyday conversations. This is achieved by using the form *shitsumon sasete itadaku* (*shitsumon sasete itadakimasu* [HUM-POL] in the formal inflection style used by the judge) for the verb *shitsumon suru* ('to ask,' literally, 'to make a question'). This highly honorific form can be translated more directly as 'I will humbly allow myself to ask you some questions.' However, because this style is relatively common in everyday speech, such a rendition would give it an excessively polite or even submissive tone, which would not reflect the true pragmatic nature of the utterance.

It is impossible to state with any degree of certainty which factors did or did not play a role in determining the honorific style used by the judge. Was it her gender, her age, the gender and the age of the defendants, or even all these factors? Would the style have been the same or different had the judge been male? This seems plausible but not guaranteed because men, too, can and do use such a level of formality and politeness in their speech patterns. In analyzing Japanese honorific style, therefore, we must be careful not to overemphasize the role of any one feature, such as age or gender, in determining how formal or casual speakers are in their interactions with others.

This leads to the conclusion that no one key factor determines which honorific style is appropriate in a given situation. More importantly for our discussion, there is no prescription for court interpreters that would give them clear instructions on how to render the Japanese honorific language into English or, in contrast, that would determine which honorific strategies to apply when translating English utterances into Japanese. The difficulties associated with this type of translation will be discussed in the subsequent section.

## 3 Difficulties in translating honorifics and translation traps

The description in Section 2 illustrated that English and Japanese honorifics differ considerably. Furthermore, it demonstrated that *keigo* is a robust and complex linguistic system whose usage depends entirely on the circumstances surrounding the conversational situation.

Thus, achieving equivalence in rendering honorifics poses substantial challenges. Huszcza (2006, 27) notes that "[f]or translation studies, the issue of rendering honorific expressions is crucial, because it is in this field that one can sometimes observe elemental translation errors, the main cause of which are the immense differences in the grammar of honorifics and the rules of linguistic etiquette between the languages."[3]

Such translation errors can occur regardless of the direction of the translational act. In the same work, Huszcza (2006, 126) observes that the fact that the English second-person pronoun 'you' can be used in both casual and formal situations may lead to a 'false linguistic consciousness,' wherein one may be tempted to believe that English speech styles should be rendered casually into other languages, regardless of the actual circumstances of the communication act. As a consequence of this misconception, the English 'you' and the verbs it accompanies tend to be rendered in the 'T' pronoun-style, even in situations where the 'V' forms are more appropriate in languages that make such a distinction.

---

3 The original quotation in Polish reads: "Dla translatoryki problem przekładu wyrażeń honoryfikatywnych jest niezwykle istotny, gdyż właśnie w tej dziedzinie można najczęściej zaobserwować elementarne niekiedy błędy tłumaczeniowe, których głównym źródłem są ogromne rozbieżności pod względem gramatyki honoryfikatywności i reguł etyki językowej występujące między językami (Huszcza 2006, 27)."

Similarly, Kiaer and Cagan address the issue of insufficient attention given to pragmatic features in translating East Asian languages, such as Korean and Japanese, into English:

> (. . .) Japanese and Korean speakers often deem the pragmatic information presented in those languages as important. It is therefore no surprise, with the explosion of prestige attracted by, for instance, Korean media such as *Parasite* and *Squid Game*, that much of the criticism one sees from source language speakers concerning translation into English concerns an impression that pragmatic meanings (relating to such items such as address terms) have been inadequately captured (Kiaer/Cagan 2023, xvi).

These observations are crucial for our discussion on rendering verbs and the honorific value they manifest from and into Japanese. It seems that translators of English into Japanese are neither immune from falling into this translation trap. This is especially visible in the translation practices of TV series and films from English into Japanese. This seemingly unrelated genre of translation is a helpful illustration of the difficulties in rendering honorifics and the 'false linguistic consciousness' leading translators to render English casually, even when arguably a more formal style is appropriate.

The following examples, which derive from the Netflix series *The Crown* (Wilson et al. 2020) about the British monarchy, demonstrate this issue well. The excerpts below come from an episode (Season 4, Episode 5) in which a working-class man named Michael Fagan, frustrated with the economic situation in Thatcher-era Britain, trespasses into Buckingham Palace and enters the Queen's bedroom in the morning. He addresses the Queen directly in the following manner (the author's back translations into English are provided in the parentheses).

Example 6
Fagan: You should hire me [to fix the paint on the palace walls].
Netflix Subtitles: *Ore ni naosasero* (VUL).
(Back translation: 'Let me fix [VUL] it.')

Example 7
Fagan: Save us all, from her [Thatcher].
Netflix Subtitles: *Ano onna kara kuni wo sukue* (VUL).
(Back translation: 'Save [VUL] the country from that woman.')

What is striking about these Japanese subtitles is that not only do they not use respectful or humble language, but no polite forms (formal inflection in Table 1) at all. Furthermore, the forms used in these subtitles can even be considered derogatory or vulgar (VUL). Such a speech style would be unimaginable should the scene take place in the Japanese context. This is because using honorific forms toward the monarch and the royal family in Japan is "absolute" for all speakers (Huszcza 2006, 37). More precisely, even if the Japanese Emperor or the Imperial

Family are not the direct addressees but merely the protagonists of the utterance, honorific forms are expected even when used in their informal inflection style in casual conversations (Marszalenko 2023), a situation similar to that discussed in the example of the student talking about 'Professor Tanaka' to their classmate (see Subsection 2.1).

When the context undergoing the translation process is non-Japanese, as in the examples above, the rendered style tends to be highly casual or, in this case, even vulgar (VUL). All the verbs in the examples are used in their informal inflection. Moreover, these verbs are in the highly casual imperative form (*naosasero* in Example 6 and *sukue* in Example 7, respectively). It must be noted here that the same forms could be considered simply casual, familiar, or even friendly rather than vulgar or derogatory if the interactants of the conversational act were, for example, close friends or family members.

Granted, the English source text also lacks any high honorifics. Still, such expressions in English can be treated at least as honorifically neutral. In contrast, as mentioned previously, in spoken Japanese, there is always a specific honorific value attributed to verbs, which can be positive or negative. Thus, the translations shown above give the impression that Fagan is talking to the Queen not as the monarch (i.e., someone at the top of the social hierarchy) but as his equal at most or even someone inferior to himself in status. Even if one considers his anger and frustration with the situation in the country and his own life, given his *tachiba* in the communication act, the choice of such a casual style in the target text is highly questionable.

Because many individuals are involved in subtitle translation (Munday 2008), the blame for such errors or misguided translation strategies lies not with the translator alone. However, the fact that this rendition gives an excessively casual (vulgar) impression of the interaction is indisputable. This may have implications for how the recipients of such interactions perceive the speaker of the source text. Subtitling TV shows or films in this way may generate erroneous impressions or harmful stereotypes about the source text culture or society (e.g., "In Britain, people speak to the royals in a disrespectful manner," or even something as absurd as "English speakers don't use honorifics").

In the context of legal interpreting, however, the consequences of such errors can be more serious. This is because such an erroneous perception may play a vital role in assessing the degree of the defendant's remorse or repentance or in determining their fate. Importantly, because court interpreters are not granted the same degree of freedom or creativity in their rendition as subtitle or literature translators, they face the challenge of how to produce target texts equivalent to the source text not only on the superficial linguistic level but also a more pragmatic one, namely in the honorific style intended by the source text speaker.

In the subsequent parts of the article, we will first investigate honorific styles chosen by court interpreters when rendering English into Japanese (Section 4), followed by a discussion on strategies they employ in dealing with honorifics when Japanese is the source language (Section 5).

## 4 Findings I: English as the source language

As demonstrated in the preceding parts of this article, the inflection of verbs plays a crucial role in either elevating (RES) or lowering (HUM) the honorific value of an utterance in spoken Japanese. This differs considerably from politeness strategies applied in English. We also observed that court interpreters are bound by limitations, which prohibit them from taking a more liberal stance toward the source text they work with. This implies that adding to or removing words or phrases from the target text to achieve an equivalent effect to that of the source text would run the risk of violating such rigid limitations. The challenge for court interpreters working with Japanese and English, however, differs depending on which of these languages is the source- and which is the target language.

When defendants or witnesses give testimony in English, the honorific value of their utterances may be considered neutral in some cases (as in the example 'Today is Saturday' (Huszcza 2006, 155) discussed in Section 2). In contrast, as was alluded to previously, such a neutral honorific value in spoken Japanese is virtually non-existent, and all utterances are either informal/casual (CAS) or formal/polite (POL). In the case of the formal inflection style, an utterance can be "merely polite" or honorific (i.e., humble or respectful).

This means that court interpreters face a dilemma in rendering such English utterances by defendants or witnesses into Japanese because they must decide whether to translate such source texts using the polite or the honorific inflection forms (i.e., those with the HUM or RES markers). As mentioned previously, the casual inflection style is reserved for interactions with family, friends, and other individuals one is close with; thus, such forms are not likely to be encountered in the courtroom (this may not be the case, however, should a witness or the defendant be a minor, but such cases will not be discussed in detail here).

In the discussion that follows, examples of renditions by court interpreters are divided into three types: 1) explanations by the defendant or a witness regarding the events under examination in the trial (4.1), 2) statements addressing the questioner directly (4.2), and 3) expressions of apology, remorse, or regret (4.3).

## 4.1 Referrals to the 'External Reality'

'External reality' (Hale and Gibbons 1999) refers to the events under examination in a trial. These events took place prior to the trial, and their veracity is what the Court attempts to verify. This is opposed to the 'courtroom reality,' which deals with what occurs during the trial.

This type of utterances is one that, in theory, should require the honorific style least. This is because the events being referred to do not address the interactants (trial participants) directly, and there is no need to express excessive respect for past events. Thus, the polite form (formal inflection) style can be considered sufficient in such utterances. Although this usually is the case, the gathered trial observation data demonstrate that occasionally court interpreters deviate from this default style and shift into more honorific forms, as in the examples below (the following abbreviations will be used in the subsequent discussion: D = Defendant, W = Witness, I = Interpreter, and J = Judge).

> Example 8
> D: It didn't show that the shipment cleared customs in the tracking number, so I called [FedEx] again [to inquire what was going on with the shipment].
> I: *Tsuiseki bangō wo tsukatte, tsūkan shita to hyōji sarenakatta no de, mō ichido denwa wo itashimasita* (HUM-POL).
> (Back translation: '[I] used the tracking number [but it] didn't show that [the shipment] had cleared the customs, so I made a phone call [HUM-POL] again.')
>
> Example 9
> W: I heard this from Michael.[4]
> I: *Maikeru kara kiite orimashita* (HUM-POL).
> (Back translation: 'I heard [HUM-POL] [this] from Michael.')
>
> Example 10
> W: I did not use such words.
> I: *Sō iu kotoba wo tsukatte orimasen* (HUM-POL).
> (Back translation: 'I did not use [HUM-POL] such words.')

Examples 8 through 10 derive from two drug-smuggling cases. In Example 8, the defendant explains how he kept track of packages (allegedly containing illicit drugs) shipped from the United States to Japan. The target text uses formal inflection rather than informal, which is customary in this formal setting. The interpreter goes one step further, however, by using the honorific (humble) auxiliary verb *itasu* ('to do' or 'to make') in the past tense (*itashimashita*).

---

[4] All names of persons provided in the examples in this article were altered.

Target texts in Examples 9 and 10 were produced by the same interpreter in a case where a witness testified against a Japanese defendant, his alleged accomplice in a drug-smuggling scheme. In terms of verb inflection, the target texts are similar because both use the verbs in the honorific (humble) continuous forms (*kiite orimashita* in Example 9 and *tsukatte orimasen* in Example 10), which are more natural in this case in Japanese than non-continuous forms would be.

The same source texts could have been rendered using the formal inflection style without the humble verb forms. The reasons why these more honorific forms were chosen are known only to the interpreters. We can, however, suggest certain possible explanations.

One such explanation is that in both cases, the defendant and the witness attempted to explain their actions to the Court. One could expect the defendant to try to present himself from a humble and cooperative side, which could be a justification for the interpreter's choice. The witness, in contrast, is not under trial; therefore, there is no explicit need for him to present himself in such a manner. However, in this particular trial, the witness was the defendant's alleged accomplice. Moreover, he had previously been tried and convicted for his role in the drug-smuggling scheme. Although he had nothing to gain from appearing humble to the Court (he had already been released from custody at the time of the defendant's trial), as the Prosecution's witness, he might have wanted the Court to find his testimony credible and being humble or polite could be a means undertaken by the interpreter to achieve this goal.

There are other possibilities, too. For example, the decisions made by the interpreters were not strategic but instead reflected the interpreters' individual style of speaking Japanese or were a manifestation of their wish to preserve the decorum of the courtroom discourse by expressing politeness and respect.

Regardless of the reason (or reasons) for the interpreters' choices, the honorifically neutral verbs in the English source texts gain an honorific value (HUM or RES) in the Japanese renditions. It must be noted, however, that such a manner of speaking Japanese does not violate sociolinguistic norms in Japan. This style can also be encountered in everyday conversations without indicating excessive humility or respect. As suggested above, such a style can simply be a way to preserve the decorum of the communication act.

## 4.2 Referrals to the 'Courtroom Reality'

We now turn to the 'courtroom reality' (Hale and Gibbons 1999), in which the source text refers directly to what is happening in the courtroom. In these utterances, the speaker addresses other participants of the trial, speaking directly about the procedures taking place in the hearing, for example, a question.

Witness testimony in a criminal trial starts with the witness taking an oath to 'speak the truth, not to conceal anything and not to speak falsely' (Hōsōkai 2011, 73; original translation). The judge then informs the witness that the failure to abide by the oath may lead to the charge of perjury. Before the witness's testimony begins, the judge verifies that the witness understands the oath and its implications. The witness then needs to clearly state that they will provide truthful testimony, as in Example 11 below:

> Example 11
> J (in the Interpreter's rendition): Do you swear [that you will tell the truth and not conceal anything in your testimony]?
> W: Yes.
> I: *Sensei itashimasu* (HUM-POL).
> (Back translation: 'I swear [HUM].')

The source text in English consists of a single 'yes,' whereas the Japanese translation is composed of a verb-based utterance. This strategy by the interpreter may seem surprising as they could have rendered the witness's answer with the Japanese equivalent – *hai*. However, the usage of the English 'yes' and the Japanese *hai* differs in many ways, and interpreters tend to choose such predicate-based renditions when interpreting single 'yes' or 'no' utterances into Japanese. This is especially the case with negative tag questions. In the case of Example 11, however, using *hai* in the target text would have likely been sufficient and clear. The interpreter chose the humble verb *sensei itasu* ('to swear,' 'to pledge') in the formal inflection style (*sensei itashimasu*). This can probably be explained by what was already discussed in the preceding section regarding Examples 9 and 10, namely, that the interpreter wanted to convey the humility of the witness, as well as his readiness to provide truthful and credible testimony.

Such a decision could be deemed too intrusive, but it must be borne in mind that given the differences between honorific strategies in English and Japanese, interpreters are forced to make certain decisions based on their understanding of a witness's or a defendant's demeanor (which, of course, may lead to erroneous choices).

In the following two examples, the defendant responds to the public prosecutor's questions, which the interpreters render using a humble style.

> Example 12
> D: I can't explain the details clearly.
> I: *Shōsai wo meikaku ni o-tsutae-suru* (HUM) *koto wa dekimasen* (POL).
> (Back translation: 'I cannot [POL] explain [HUM] the details clearly.')

> Example 13
> D: I can't answer that.

> I: *Sono shitsumon ni taishite o-kotae-suru* (HUM) *koto wa dekimasen* (POL).
> (Back translation: 'I cannot [POL] answer [HUM] that question.')

The same strategy was applied in both examples. The interpreters use the humble verb form (*o-tsutae-suru* and *o-kotae-suru* in Examples 12 and 13, respectively) in their base form (which, in this case, can be considered equivalent to the English infinitive) coupled with the auxiliary verb *dekinai* ('cannot do,' 'be unable to') in its formal inflection style (*dekimasen*) as the predicate.

The interpreters' choice can be explained by the fact that when responding to the questions by their adversary – the public prosecutor – the defendant is expected to be respectful and polite and, at the same time, may wish to avoid sounding aggressive or excessively defiant. Thus, using these humble verb forms, the defendant (through the interpreter) comes off as assertive but sufficiently polite; in other words, he "respectfully disagrees."

## 4.3 Expressions of regret or remorse

Remorse is one of the most important, if not *the* most important, emotions, whose expression is expected in a Japanese criminal trial if the defendant pleads guilty to the charges. This is demonstrated by the fact that even judges during sentencing and prosecutors in their closing arguments treat expressions of remorse by the defendant as a mitigating circumstance, which may impact the punishment imposed on the defendant (the differences between 'remorse' [*hansei*] and 'regret' [*kōkai*] in English and Japanese are too complex to be addressed here, but a detailed discussion can be found in Torikai 2004).

Thus, in their chief examination of the defendant, it is common for the defense counsel to ask their client questions such as 'How do you feel about your actions now?' This signals to the defendant that they should now "take responsibility" for the consequences of the crime by expressing remorse. Several examples of defendants responding to such a signal are presented below.

> Example 14
> D: Very regretful.
> I: *Hijōni kōkai shite orimasu* (HUM-POL).
> (Back translation: 'I regret [HUM-POL] [it] very much.')

> Example 15
> D: I am so sorry.
> I: *Taihen mōshiwakenai to omotte orimasu* (HUM-POL).
> (Back translation: 'I am very sorry [HUM-POL].')

Example 16
D: It was a huge mistake, and I regret (HUM-POL) it.
I: *Ōkina ayamachi datta to omoimasu shi, kōkai shite orimasu* (HUM-POL).
(Back translation: 'I think it was a big mistake, and I regret [HUM-POL] it.')

All target texts above are similar because the interpreters use the humble continuous verb form (*-oru*) in the formal inflection style (*-orimasu*). Out of the utterances discussed in this article, using such humble forms to render the speaker's expressions of remorse into Japanese is arguably the most justified. This is because it is in their answers to such questions that the defendant has the opportunity to demonstrate their humility and acknowledgment of guilt, which can be an essential factor in deciding how severe or lenient the punishment will be.

It is impossible to predict how a judge or a judicial panel would perceive the defendant should the interpreter use the "merely polite" forms rather than the humble ones in such utterances. Neither is there any way of knowing if and how such a perception could alter the degree of the punishment. Possibly, however, the polite forms without the honorific value markers (HUM or RES) could be perceived as insufficient in this context. Therefore, the defendant's remorse could seem insincere or a pro forma device deployed merely to reduce their punishment (which, of course, in some cases, it just might be).

## 4.4 Summary

In this section, it has been shown that although the polite (POL) inflection is the default, and thus, sufficient, style of rendering English utterances into Japanese, court interpreters may and do deviate from it by coupling it with more honorific (humble or respectful) verb forms. This may be controversial because it suggests that interpreters *decide* to choose such more honorific forms, although they are not mandatory. However, a choice cannot be avoided because English lacks such a distinction (i.e., the distinction between the POL and HUM-POL/RES-POL forms). Selecting the polite forms could be considered a "safe" choice as the interpreter cannot be accused of exercising excessive discretion. On the other hand, should the interpreters aim for a more pragmatic equivalence in their rendition, these honorific forms could be a valuable tool to achieve it. Moreover, as mentioned in the discussion on expressions of remorse and regret (subsection 4.3), the "merely polite" forms may, to some recipients of the target text (i.e., judges or lay judges), seem insufficiently honorific and, as a result, the defendant insufficiently remorseful. In the subsequent part of the article, we will investigate how the court interpreter's challenge changes when Japanese is the source- and English the target language.

# 5 Findings II: Japanese as the source language

Unlike in English-to-Japanese interpreting, when Japanese is the source language, the interpreter's choices are limited because, as already discussed, in contrast to spoken Japanese, it is possible to express oneself honorifically neutrally in English. However, this, too, is not without consequences. When interpreting Japanese honorifics into English, their honorific value is usually neutralized. More specifically, the complexity and intricacies of the Japanese *keigo* style tend to be sacrificed. This is because adding words or phrases (for example, honorific titles such as 'Sir,' 'Madam,' or 'Your Honor') not explicitly present in the Japanese original to make up for the lack of honorifically marked verbs in English and to thus bring the target text closer to the source text on a more pragmatic level, could be considered excessive interventions, and so, court interpreters tend to demonstrate restraint in applying such strategies.

In the interpreter-mediated criminal trials observed for this study, verbs in the honorific style (RES or HUM) were found mainly in utterances by judges and lay judges. It is not surprising that public prosecutors or defense counsels tend not to use such a style when addressing the defendant. The prosecutor, by definition, plays an accusatorial and adversarial role vis-à-vis the defendant. In contrast, the defense counsel is the defendant's proxy and, thus, may aim to present themselves and their client humbly. On the other hand, judges and lay judges are expected to be impartial, and their use of honorifics may merely manifest the wish to preserve the courtroom decorum mentioned earlier.

In the following subsections, we will examine examples of *keigo* usage in utterances referring to the procedures taking place in the trial (5.1) and those where such honorific strategies were applied in a judge's (J) or a lay judge's (LJ) direct questions to the defendant or a witness (5.2).

## 5.1 Utterances pertaining to courtroom procedures

Instances of utterances in this category were scarce in the dataset collected for this study. One such utterance is presented in Example 17 below.

> Example 17
> J: *Hitochigai de wa nai ka wo kakunin suru tame, ikutsuka no <u>shitsumon wo sasete itadakimasu</u>* (HUM-POL).
> I: I <u>would like to ask</u> you some questions for verification purposes.
> (Back translation: 'I <u>will allow</u> [HUM-POL] myself <u>to ask</u> some questions for verification purposes.')

This standard utterance made by the presiding judge at the beginning of a trial was already discussed in detail in Example 5 in subsection 2.2.2. As already mentioned in the discussion pertaining to that example, rendering the honorific verb form *shitsumon (wo) sasete itadaku* (*shitsumon [wo] sasete itadakimasu* in the formal inflection; 'to ask') more "literally," for example, 'I will allow myself to ask you some questions,' would be overly polite compared to the pragmatic force of the Japanese original and may even be interpreted as sarcasm considering a judge's position in a trial.

Thus, given their limited tools, the interpreter chose a different strategy in this case. Rather than producing a "plain" target text devoid of any honorific value, such as 'I will ask you some questions,' the actual target text uses 'would like to,' which arguably sounds to most English speakers as more polite and, indeed, more humble, than the more definitive and assertive future tense.

Another strategy available to the interpreter could have been a target text in the vein of 'Let me ask you some questions.' This rendition would have conveyed the nuance of the Japanese form *sasete itadaku* ('to allow oneself to'). However, it would likely have scored lower on the honorific continuum than the target text in Example 17.

## 5.2 Direct referrals to the addressee and their circle

This category of utterances lies at the heart of the Japanese honorific language, *keigo*. As mentioned in Section 2, whether a speaker in a formal communication act chooses humble or respectful forms depends on who the addressee and the protagonist of the utterance are. Should the protagonist be the speaker him- or herself or those in their circle (their family, friends, or coworkers), the honorific style will likely be humble. In contrast, if the speaker refers to their addressee or those in the addressee's circle or if the protagonist is someone deserving of respect in general (for example, the Emperor or the Imperial Family, or persons worthy of respect or admiration for other reasons), the verbs will take on a respectful form.

As trial participants other than the defendant or witnesses seldom have the opportunity to speak about themselves or their circle (except in situations such as those discussed in the preceding subsection), one could expect more respectful, rather than humble, forms in their utterances in Japanese. This applies to utterances directed at defendants and witnesses as well, as in Examples 18 through 20 by judges (J) and a lay judge (JL), presented below (the role of lay judges in the Japanese legal system can be compared to juries in other jurisdictions; however, due to the differences between the Japanese system and systems in other countries, the nomenclature differs as well; see Marszalenko 2013 for details).

> Example 18
> J: *Hikokunin no okusama wa, honken ni tsuite dō omou ka wa, anata wa <u>gozonji desu</u>* (RES-POL) *ka?*
> I: Do you <u>know</u> what the defendant's wife thinks about the incident?
> (Back translation: 'Do you <u>know</u> [RES-POL] what the defendant's wife thinks about the incident?')

This question by a judge was directed at a witness in a drug-smuggling case. The witness was the defendant's former superior at work and was summoned to testify as a character witness for the Defense. The protagonist of the question is, at the same time, the direct addressee, i.e., the witness, who was a high-ranking US military official stationed in Okinawa testifying in his official capacity. Thus, the witness is clearly someone in the defendant's circle.

The judge uses the honorific (RES) verb *gozonji dearu* ('to know') in its formal (POL) inflection style (*gozonji desu*). Such an honorific style poses a challenge for the interpreter if they wish to convey the verb's honorific nuance in English. This is because, unlike in Example 17 in subsection 5.1, where the interpreter was able to use the polite 'would like,' in Example 18, it would be hard to find an honorific alternative for the English verb 'to know' without having to add words or phrases that are not present in the source text.

One such alternative, should the interpreter be allowed to exercise this much liberty, could be to use the form of address 'sir,' as in: 'Do you know, sir, what the defendant's wife thinks about the incident?' This form of address is common in military circles and thus could sound natural to the witness. However, it could be challenging to explain why a judge addresses a witness in such a respectful way. Further, it could also lead to controversy, as the concentration of the US military on the island of Okinawa is often a bone of contention between the local authorities of Okinawa Prefecture and the central government of Japan (see Nomura/Shimabuku 2012 and Nomura 2019). For this and other possible reasons, such a rendition by the interpreter would have likely been difficult to justify.

Other alternatives could also be controversial or risky, which might have been why the interpreter chose to omit the honorific value of the phrase in question in the source text altogether and render it with the honorifically neutral English verb 'to know.' This, of course, does not change the meaning of the judge's question in any way but demonstrates that certain features of Japanese honorifics cannot make it into the target text due to the limitations put on court interpreters.

In Examples 19 and 20 below, the questions by a lay judge (LJ) and a judge (J), respectively, address the defendant directly.

> Example 19
> LJ: *Hikokunin no kata ni kanojo ga irassharu* (RES) *to iu hanashi ga arimashita* (POL).
> I: You mentioned you had a girlfriend.
> (Back translation: 'There was [POL] a mention that the defendant has [RES] a girlfriend.')

Multiple honorific strategies are at work in Example 19. The defendant himself is referred to respectfully (*hikokunin no kata*, which can be rendered literally as 'the respectful accused person'). As the focus of this study is the usage of verbs, however, we will now examine how this part of speech is used to give a respectful (RES) value to the utterance.

In both the English target text and the back translation, the possessive verb 'to have' was used. However, rendering this Japanese sentence more closely would produce an English target text in the vein of 'There was a mention that there is a girlfriend to the defendant.' This is because the verb *irassharu* (here used in an infinitive-like form because it is not the predicate of the sentence) is closer to the English 'to be' rather than 'to have' (due to many differences between 'to be' and the numerous Japanese existential verbs, however, *irassharu* should not be treated as an equivalent in all situations).

Such a verb usage poses the question: At whom is the honorific value of the utterance directed? Is it the defendant or the girlfriend, or maybe both persons? This question is unlikely to be answered definitively; thus, for the reason of simplicity, we can broadly define the recipient of the respect given in the utterance as "the defendant and his circle."

> Example 20
> J: *Shazaibun wo kakareta* (RES) *no wa ichigatsu datta to iu koto desu* (POL) *ga, naze kaku koto ni natta no desu* (POL) *ka?*
> I: You said you wrote the letter of apology [to the victim] in January; why did you decide to write it?
> (Back translation: 'It was [POL] mentioned that you wrote [RES] the letter of apology in January, but what was [POL] it that led to your writing it?')

Defining the direct recipient of the honorific value in Example 20 is more straightforward than in Example 19. Here, this recipient is clearly the defendant, who is both the addressee and the protagonist of the question. The verb used in the first part of the sentence (*kakareta*) is the past tense of the honorific (RES) form (*kakareru*) of the verb *kaku* ('to write'). It is also one of the forms discussed in Table 1 (see Section 2). This verb is used in a form similar to the English infinitive because it is not the main verb (predicate) of the sentence (which is *desu* used here with the question particle *ka*). Should *kaku* (*kakareru*) be the predicate of the question, the form used would have been *kakaremashita* rather than *kakareta*. These details of Japanese syntax are not as important for our purposes, however, as the fact that the verb form used for the action of writing by the defendant is an honorific (RES) one.

It would be interesting to learn what led the judge to use this respectful style to address the defendant. From the perspective of the discussion at hand, however, it is not as important as the consequences of this style for the production of the target text. Regardless of the reason (or reasons), it is clear that preserving the honorific value of the source text in the target text would be challenging, if not impossible, as was also the case in Example 19.

## 5.3 Summary

Examples in this section have demonstrated that rendering Japanese honorifics into English usually leads to their honorific value being neutralized. There are multiple reasons for this. To begin with, English lacks the distinction between formal (polite) and informal (casual) inflection styles of verbs, which is one of the most prominent features of Japanese. As we have already discussed, in spoken Japanese, speakers have *no choice but to choose* either of the two styles, thus making honorifically neutral utterances virtually impossible. Furthermore, certain verbs in Japanese have an inherent honorific value, regardless of the inflection style (e.g., *gozonji dearu* or *irassharu* discussed in subsection 5.2). The difference between these verbs and their more neutral counterparts (neutral only in their basic, infinitive-like inflection forms) is another aspect of the Japanese language difficult to render into English. Should interpreters wish to convey the pragmatic aspect and nuances of this Japanese honorific style, they would most likely need to exercise a higher degree of liberty toward the source text. Such strategies in court interpreting are usually not permitted and could lead to far-reaching consequences concerning the perceived accuracy and thus the perceived reliability of the interpreting services provided.

# 6 Concluding remarks

This article has addressed the challenges and difficulties in translating honorific language between English and Japanese. It has demonstrated that honorific strategies applied in both languages differ considerably and that rendering these strategies into the other language rarely, if ever, produces "literal" or "word-for-word" translation. We have seen that when court interpreters deal with the Japanese honorific language – *keigo* – its robust and complex nature is usually sacrificed in the English translation or offers the court interpreter multiple and sometimes unavoidable choices when Japanese is the target language.

Furthermore, we have discussed that honorifically neutral utterances in spoken Japanese are all but impossible, and any utterance that contains a predicate must be either casual or polite in terms of verb inflection. When rendering a Japanese utterance into English, most features of the complex Japanese honorific system are sacrificed because English lacks the distinctions between casual and formal inflections and is not as rich in terms of verbs with inherent honorific values as Japanese. Conversely, when English is the source language, the court interpreter needs to decide which honorific style is more appropriate in each situation: "merely polite" (i.e., polite (POL) forms without the RES or HUM markers) or more honorific (humble or respectful). These decisions are based on multiple factors, such as the speaker, the addressee, the speaker's demeanor, or the intent and nature of the utterance. This, of course, poses the risk of an erroneous or controversial choice by the interpreter.

In the author's view, all the examples of renditions presented in this article are within the "range of correct translation" alluded to by the interviewed court interpreter (Marszalenko 2016, 37) quoted in the Introduction. As evaluating accuracy can be quite subjective, however, judging whether or how much court interpreters should deviate from the default "merely polite" style (i.e., formal inflection without the honorific markers) by using the more honorific (humble or respectful) forms depends on how one defines accuracy in the context of interpreting in criminal trials. Should accuracy be understood in more pragmatic terms, certain choices by court interpreters may be not only justifiable but also desirable. In contrast, should one deem such renditions by court interpreters excessive, one would need to conclude that the differences between honorific styles in Japanese are not sufficiently significant to be reflected in the interpreting process and that they can be sacrificed or ignored.

Regardless of one's stance toward accuracy, however, the discussion in this article demonstrates that "word-for-word" translation is not possible when honorific strategies, especially those in Japanese, are applied. It also clarifies that the choices made by court interpreters may be informed by a variety of factors, not all of which may be made as part of entirely conscious cognitive processes. Moreover, the Japanese honorific language may be used merely to preserve the decorum of the formal courtroom discourse rather than manifest deep humility or respect, making court interpreting even more challenging.

The findings presented in this article are derived from a limited number of interpreter-mediated criminal trials heard within a limited period. Thus, they cannot be treated as the representative styles of all or even most court interpreters working with English in Japan. They do, however, shed light on the challenges court interpreters face in the provision of their services and allow us to advance the discussion on the issues of accuracy and equivalence in the context of court interpreting. Further research into this topic will undoubtedly provide even more insights into these matters, thus deepening our understanding of the work of court interpreters. This, in turn, will positively impact access to justice for speakers of languages other than the official language of the judicial system in the jurisdiction in question.

# Bibliography

Hale, Sandra & John Gibbons. 1999. *Varying Realities: Pattern Changes in the Interpreter's Representation of Courtroom and External Realities*, Applied Linguistics 20/2, 203–220.
Hōsōkai. 2011. *Hōtei Tsūyaku Handobukku* [Courtroom Interpreting Handbook], Tōkyō, Hōsōkai.
Huszcza, Romuald. 2006. *Honoryfikatywność* [Honorifics], Warszawa, Wydawnictwo Naukowe PWN.
Inoue, Miyako. 2003. *Vicarious Language. Gender and Linguistic Modernity in Japan*, London, University of California Press.
Kadrić, Mira. 2021. *Legal Interpreting and Social Discourse*. In Meng Ji & Sara Laviosa (eds.), *The Oxford Handbook of Translation and Social Practices*, New York, Oxford University Press, 501–520.
Kiaer, Jieun & Ben Cagan. 2023. *Pragmatics in Korean and Japanese Translation*, New York, Routledge.
Laster, Kathy & Veronica Taylor. 1994. *Interpreters and the Legal Process*, Leichhardt, The Federation Press.
Marszalenko, Jakub E. 2013. *Three stages of interpreting in Japan's criminal process*, Language and Law/ Linguagem e Direito 1(1), 174–1187.
Marszalenko, Jakub E. 2016. *Conduits, communication facilitators and conduits: Revisiting the role of the court interpreter in the Japanese context*, SKASE Journal of Translation and Interpretation 9(2), 29–44.
Marszalenko, Jakub E. 2023. *Looking for Clues in Most Unlikely Places: What the media discourse on Japan's Imperial Family can tell us about challenges in Japanese-English court interpreting*, Bulletin of Nagoya University of Foreign Studies 12, 51–84.
Munday, Jeremy. 2008. *Introducing Translation Studies*, New York, Routledge.
Nomura, Kōya & Annmaria Shimabuku. 2012. *Undying Colonialism: A Case Study of the Japanese Colonizer*, The New Centennial Review 12(1), 93–116.
Nomura, Kōya. 2019. *Muishiki no Shokuminchishugi – Nihonjin no Beigun Kichi to Okinawajin* [Unconscious colonialism: Japan's US military bases and the Okinawans], Tōkyō, Shōraisha.
Obana, Yasuko. 2020. *Politeness*. In Patrick Henrich & Yumiko Ohara (eds.), *Routledge Handbook of Japanese Sociolinguistics*, New York, Routledge, 248–263.
Shimotani, Maki. 2012. *Shizen danwa ni okeru nininshō daimeishi 'anata' ni tsuite no ichikōsatsu- ninshiteki yūsei (epistemic primacy) wo fumaete* [A study on the second-person pronoun 'anata' in natural discourse: the issue of epistemic primacy], Papers in Teaching Japanese as a Foreign Language 22, 63–96.
Torikai, Kumiko. 2004. *Rekishi wo kaeta goyaku* [Mistranslations that changed the world], Tōkyō, Shinchōsha.
Wilson, Jonathan D. & Peter Morgan (Writers), Paul Whittington (Director). 2020, November 15), *Fagan* (Season 4, Episode 5). In Peter Morgan (Executive Producer), *The Crown*, Netflix.
Yonezawa, Yoko. 2021. *The Mysterious Address Term anata 'you' in Japanese*, Amsterdam/Philadelphia, John Benjamins Publishing Company.

Caroline Laske
# Textual representation as a conceptual tool: Big data analysis of legal language

**Abstract:** This paper argues for a sociolinguistic approach to legal linguistics, which contributes insights into the law that have hitherto been under-explored, in particular, in legal scholarship. Law as a tool of social governance is conditioned by the society's perceptions of its identity, including the hegemony of its structures, its moral, social, geo-political and cultural constructs and its economic needs. This aspect of social governance is linguistically fixed through legal language, which is performative, on the one hand, but also conveys social meaning, sociolinguistic content and aspects of the society's identity, on the other. That aspect of legal language has had less attention in legal research.

**Keywords:** legal linguistics, socio-linguistics, pragmatics, textual representation, gender

## 1 Introduction

Over the last few decades, much of the interdisciplinary thinking between law and linguistics has revolved around issues that resulted from the specialised and abstract nature of legal language: subjects such as terminology & legal concepts, translation into other languages, into other legal systems or on an international level or the inaccessibility of legal language (LSP v general language) tended to be at the heart of legal linguistics research. Yet, while a number of linguistic variants used in specific professional settings relate to niche matters with a relative small community of language usages, legal language, despite its speciality status, cannot be perceived in such a way. Legal language is in wide-spread use describing law, justice and the legal system, all phenomena that govern every aspect of our lives. The all-pervasiveness of law lies in our constant contact with it, both on a personal and societal level. Law as a tool of social governance is conditioned by the

---

**Note:** The research for the studies described in section 3.3 was done during the author's stays as Heinz Heinen Fellow at the Bonn Centre for Dependencies and Slavery Studies (2019–2021) and as CAS Fellow at the Centre for Advanced Studies, Norwegian Academy of Science and Letters (2021–2022).

---

**Caroline Laske,** University of Louvain

https://doi.org/10.1515/9783110799651-008

society's perceptions of its identity, including the hegemony of its structures, its moral, social, geo–political and cultural constructs and its economic needs. This aspect of social governance is linguistically fixed through legal language, which is performative, on the one hand, but also conveys social meaning, sociolinguistic content and aspects of the society's identity, on the other. That aspect of legal language has had less attention in legal research.

Our perception of legal language as a tightly knit, highly technical jargon with specialised terminology and a formulaic character is in part grounded in its performative function. Matilla identifies legal language as having the function of transmitting legal messages, strengthening the authority of the law and maintaining order in society, reinforcing the team spirit of the legal profession and, lastly, linguistic policy goals (Matilla 2013, 41–86). The law and its practices have realised acts through the use of language. This is grounded in the law's historical traditions, its normative nature, prescriptive and performative functions and the basic premise that the letter of the law is supreme. To that extent it is one of the quintessential areas where the theory speech acts of Austin and Searle reveals all its significance. Law, as we know it in the Western world, usually comes in written form. Indeed, in many jurisdictions the performative aspect of law requires the text to be written and published for it to become valid law and produce legal effect. The intrinsic link between law and language in Western European culture means that law can no longer be imagined without, in particular, the use of written language. 'Words are the lawyer's tools of trade', Lord Denning, English judge and Master of the Rolls (head of civil justice) famously wrote (Denning 1979, 5). Legal language fulfils several functions, the most important of which is probably to achieve justice by 'producing legal effects by speech acts' (Matilla 2013, 41).

Over the last few decades, the linguistic turn has been integrated into legal scholarship, but to a different degree than in other disciplines. The importance of studying legal language can be measured in the rising number of books, including comprehensive handbooks (Matilla 2013; Freeman/Smith 2013; Tiersma/Solan 2016; Bhati 2012; Baaij 2012; Ramos 2013; Giannoni/Frade 2010; Coulthard/Johnson 2012) and specialised periodicals[1] covering, among other things, legal communication, harmonisation of laws (EU), legal translation, semantics, terminology, and forensic linguistics. Scholars have also used methodologies from computational linguistic disciplines in the study of law and legal language, mainly in relation to questions of translation of EU law and of other supra- national/international law (Mori 2018;

---

[1] See, e.g. International Journal of Language & Law, Zeitschrift für Europäische Rechtslinguistik, International Journal of Legal Discourse, Comparative Legilinguistics, International Journal for the Semiotics of Law.

Biel/Doczekalska 2020). But the study of legal text as the linguistic expression of law using language as a naturally occurring phenomenon with inherent discourses and biases has less prominence in legal scholarship. Social/political scientists and linguists have given some attention to biased discourses in law and law-making, lawyers and legal historians may have studied the effect of laws and law enforcement in relation to specific issues of repression or constraints. But in legal scholarship, relatively little systematic analysis of legal language has examined it as a naturally occurring linguistic phenomenon *per se* under the microscope for its underlying meanings, semantic changes and discourses. This paper argues for the use of corpus linguistics (CL), combined with discourse analysis and a sociolinguistic approach, to provide contexts that enable us to gain deep-level understanding of the origins, evolution and change of legal thinking and the law by studying the language in which they are expressed. It must be stressed that CL methodologies are considered here as *tools* to reveal *tendencies* and patterns in legal language usage and ways of encoding meanings, and the resulting data contributes to specific legal research questions. It is therefore essential that the use of such tools is preceded by specific research questions – the linguistic study and the data then contribute to the overall research. The discussion in this paper illustrates how CL can be a valuable tool in a variety of areas of the law, research questions and historical periods.

The paper falls into two main parts. The discussion that follows will retrace some of the reasons why legal language has become an intricate jargon of highly technical and abstract terminology, which explains in large parts why interdisciplinary thinking and research between law and linguistics have mainly dealt with issues arising from that aspect of the phenomenon of legal language. In the second part of the paper, legal language will be discussed from a pragmatic angle and in relation to its underlying sociolinguistics content. Three scenarios will be described: (i) the use of CL in the search for ordinary meanings of legal language in the context of the interpretation of statutes and judicial decisions; (ii) the concept of semantic objectivity as the result of linguistic convention; (iii) the textual representation in legislation of specific categories of persons and how the social perception of their legal rights/capacity can be revealed by studying the underlying meanings of the language in which they are written. The research discussed in this third scenario concern thirteenth-century women in Normandy, Saxony and Norway and the concordance-based studies reveal the social perception of them as a group beyond the content analysis of legal provisions.

## 2 The phenomenon of legal language

### 2.1 Performative characteristics

Historically, the performative aspect of exercising power tended to be ceremonial and formulaic, emphasising the authoritative nature of the situation. This applied well before laws and legal texts were written. Even today, the set-up we find in many courts of justice, such as the wearing of gowns and, in some jurisdictions, of wigs, having to rise when the judges enter etc., are a part of the non-verbal performative aspects of law. The use of written language added another layer to this performative nature. A written document made a subject matter less ephemeral and of continued relevance for future purposes. Moreover, in largely illiterate societies where writing was associated with the learned and powerful, a written legal document reinforced the exercise of legal and political power. Written (legal) texts were also presented as having been of divine inspiration. For example, in the prologues to both the Saxon *Sachsenspiegel*[2] and the Norwegian *Landslǫg Magnúss Hákonarsonar*[3] there are direct references to God/Holy Spirit as divine inspirations for the texts.

The performative character of law was retained through ceremonial procedures and formulaic language that was written down and therewith reinforcing its official and authoritative character. This performative nature of legal language also meant it became specific and separate from everyday language with its increasingly specialised terminology, describing ever increasingly complex legal concepts. The combination of highly technical and specialised subject matter and terminology with its specifically performative characteristics resulted in a technical jargon increasingly separated from everyday usages of language. As discussed below, the use of a language/jargon that became totally incomprehensible to anyone but the initiated, also had the effect of keeping law–making and practices among the elites who had the knowledge of the specialised language used, which in itself reinforced the isolation of the language from everyday uses. It is against

---

[2] "Des hiligen geistes minne, diu sterke mine sinne, dat ek recht unde unrecht der Sassen bescede na Goddes hulden unde na der werlde vromen . . . Got is selve recht . . ." (*Sachsenspiegel*, Prologus) ["May the love of the Holy Spirit sharpen my mind so that I may pronounce what is lawful and unlawful among the Saxons for the grace of God and the benefit of the world . . . God is law itself . . ." (translation Maria Dobozy)].

[3] "Magnús, með guðs miskunn Noregs konungr, son Hákonar konungs, sonarson Sverris konungs, sendir ǫllum guðs vinum ok sínum í Frostuþingslǫgum kveðju guðs ok sína." (*Landslǫg Magnúss Hákonarsonar*, Prologus) ["Magnus, the King of Norway by the grace of God, son of King Hakon, grandson of King Sverrir, sends God's and his own greeting to all friends of God and his in the Gulathing district" (translation Jóhanna Katrin Friðriksdóttir)].

that backdrop of highly technical terminology and specialised language, coined today as language for special purposes or LSP, that is at the heart of the bulk of legal linguistics research.

## 2.2 Language for special purposes

During the Medieval period, when it became the practice to use writing in law and justice, Latin as the learned language, offering terminology and suitable legal formulae from Roman law, played an essential part (Mattila 2013, 161–201). But while Latin was relatively widespread, the so-called Law French, a French dialect specifically used in the common law of medieval England, was restricted to the very inner circle of the common law justice system (Laske 2016; 2018). In its beginnings, the French used in the English common law was not as such a technical language. Its origins went back to the Norman conquest of England and the subsequent reign of the French–speaking Norman and Plantagenet kings. In the second half of the 13th century,[4] French became the language used for some official documents, legal tracts, treatises and statutes (though writs, plea rolls and other official records remained in Latin) – "something happens to make Englishmen write about law in French and frame statutes in that language" (Woodbine 1943, 402). But with time and the general decline of French in England, Law French became a highly specialised use of that language within the closed confines of the legal profession and abstract from everyday usages. The language contact was such that, on the one hand, the French used in the law took an independent path from the general French that became widely absorbed into the English language. On the other hand, lawyers speaking native English and acquiring French, constituted the link that allowed for great quantities of French words with legal connotations to penetrate the English language (Mellinkoff 1963, 109).

During the 16th century, Law French was still used at the Inns of Court and very occasionally in the law courts. The early law reports[5] were also written in French. The legal profession hung onto this idiom that had shaped their law, legal thinking, habits and the construction of their concepts and arguments.[6] Coke described Law French as: "vocabula artis [. . .] so apt and significant to express the true sense of the laws, and are so woven in the laws themselves, as it is in a manner impossible to change them [. . .]" It was an idiom known to the noble and

---

4 There is textual evidence that French was used in legal texts before the second half of the 13th century, see Rothwell 1975, 457; Rothwell 1993, 262.
5 The so-called 'old' reports first appeared during the reign of Henry VIII and ran until 1865.
6 For example: fee simple was fee simple, fee taile became fee tail, heires became heirs.

wealthy classes and their sons, often educated in the law. Even from the time when English became more common, the wealthy, keen to maintain their privileges, through land law in particular, continued the use of Law French for the reasons mentioned above, but also to "lock up trade secrets in the safe of an unknown tongue" (Mellinkoff 1963, 101). John Warr[7] suggests that: "the unknownness of the law, being in a strange tongue; whereas, when the law was in a known language, as before the Conquest, a man might be his own advocate. But the hiddenness of the law, together with the fallacies and doubts thereof, render us in a posture unable to extricate ourselves; but we must have recourse to the shrine of the lawyer, whose oracle is in such request, because it pretends to resolve doubts" (Warr 1650/1810, 221–223). There is little evidence for this being a deliberate process, but it was certainly a collateral benefit for the powerful (Ormrod 2003, 765). The story of Law French is an excellent almost extreme example of how legal language can become separated from everyday tongues to a point where it is actually a different dialect only mastered by those initiated in the law and how that also allows them to perpetuate their power and influence in a situation akin to a closed shop.

Despite formulaic characteristics that were originally conditioned by the use of Latin and the performative usages, the textual evidence shows that the early vernacular legal languages were not just technical jargon but also showed general language register that subsequently grew into a highly specialized language. As societies developed, different socio-political and economic needs shaped law and legislation. For example, as commercial trade was growing and widening, also in geographical terms, new legal concepts and provisions were needed in contract law or in relation to debt. The rising complexity of societies and socio–economic relations were mirrored in more complex legal concepts and increased technicality of the legal language. This often went hand in hand with neologisms and semantic shifts of existing terminology, which were not necessarily linear but could be haphazard and even arbitrary at times. A recent study (Laske 2020[a]) on the legal enforceability of informal agreements/contracts in late medieval English common law has shown that the shift from earlier proprietary notions of exchange to the emerging concept of promissory undertaking was not reflected in the terminology used. The linguistic data shows no definite new legal technical terms or categories, nor can we observe uniformity in the terminology of the legal actions. Instead the language used terminology that linguistically encoded

---

7 John Warr was a Leveller and independent reformer of law, arguing for the participatory liberty of the individual as the basis of all legal norms. Levellers and reformers like him initiated the process of establishing individual liberty at the centre stage of the political process. Later this was fully developed by Locke and the modern natural law tradition.

old established categories of proprietary concepts. The fact that the terminology lagged behind the development of informal contracts from a conceptual point of view highlights the concordance between the complexity and difficulties to establish this new concept and to find appropriate terminology. Furthermore, once the use of new terminology stabilized, a diachronic corpus study highlighted the semantic shifts away from general language use to increased terminological abstraction alongside the consolidation and development of the new legal thinking. The lexical variation of the textual contexts became increasingly poor and semantically restricted, which is indicative of a process of increased abstraction and of specialized language.

# 3 The need for pragmatics and sociolinguistics

Besides these formulaic and highly specialised aspects of legal language discussed so far and with which a great deal of the legal linguistics research has been concerned, legal scholarship has paid less attention to more sociolinguistic aspects of the use of language in the law and in particular approaches based on pragmatics and on big data analysis. To that extent, legal research is missing out on insights derived from additional contextual data and information that these linguistic approaches can provide. For example, textual representation is a concept rarely used in legal research. While language and terminology have always been intrinsically linked to the existence of written law, the textual representation with encoded meaning, discourses and biases is not very prominent in legal research. The insights and additional contexts that such an approach can offer tend to be less part of research questions in legal scholarship. In this section of the paper, we will discuss a number of recent studies that have concentrated on legal language as language in use and less as one for special purposes.

## 3.1 Legal interpretation

The interpretation of language in the context of statutory interpretation and judicial decisions frequently requires the search for the ordinary meanings of terms that appear in legal language. Traditionally that has involved the use of tools such as dictionaries, etymology, as well as judicial intuition. Yet, these restrict the search for semantic meanings. While a dictionary usually lists several senses of a word and includes the context in which they occur, it is still a synthesis and the result of several choices (inclusion/exclusion, categories, etc.) made during compilation. But the formation of meanings is conditioned by the contexts in which they occur and therefore goes beyond abstracted semantics of dictionaries (Recanati 2004, 5–22). The pragmatic view of language considers the formation of meanings not only as being dependent on the speaker/listener's structural and linguistic knowledge but also on contextual elements such as the cultural and situational context, intertextual knowledge, inferred intents. John Firth (1957, 11) expressed this context-dependent nature of meaning in his seminal phrase: "You shall know a word by the company it keeps" though that is a reference to the specific linguistic context. Yet, if we consider language as an act of communication, context alone cannot fully account for how meaning is affected. Relevant contexts are determined by the function of what is said or written, and what information is to be communicated (Halliday 1976; 1985; Halliday/Hasan 1985). The contextual element to meaning is particularly relevant in relation to the understanding of specialised language, such as legal language. Hence, a differentiation needs to be made between specialised meanings and ordinary language meanings. A big data approach involving corpora of naturally occurring texts will offer considerably more quantitative and contextual information on pragmatic meanings and uses in naturally occurring language environments. In the case of diachronic corpora, it also allows us to trace the evolution of language usages and (shifts in) meanings historically, which, in law, can represent several decades or even centuries between the drafting of legislation and its application in the courts.

The use of corpus linguistics in the context of legal interpretation has been championed by some judges on different US state supreme courts.[8] Federal courts of appeals,[9] practicing lawyers[10] and scholars (Solan 2016, 59; Solan/Gales 2017, 1311) in the US have discussed the contribution that CL can bring to the interpretation of legal language. Lee/Mouritsen (2018) argue for the use of corpus linguistics tools and how these methodologies can yield empirical data to questions of

---

[8] See, e.g. *State v Lantis*, 447 P3d 875, at 880–81 (Idaho 2019); *People v Harris*, 885 NW2d 832, at 838–39 (Mich 2016); *Richard v Cox*, 2019 UT 57, 14–25, 450 P3d 1074, at 1078–81; *State v Rasabout*, 356 P3d 1258 (Utah 2015). See Lee/Mouritsen 2021.
[9] See, e.g. *Wilson v Safelite Group, Inc.*, 930 F3d 429, 139, 444 (6th Cir 2019): Thapar, J suggesting the use of CL as an 'important tool' in the 'judicial toolkit', reliance on searches in the Corpus of Historical American English; *Ceasars Entertainment Corp. v Int'l Union of Operating Engineers Local 68 Pension Fund*, 932 F3d 91, 95–96 (3rd Cir 2019): use of CL in conjunction with dictionary definitions. See Lee/Mouritsen 2018.
[10] The 'emoluments' litigation drew substantial attention from linguistics scholars, e.g. Brief of Prof C Cunningham and Prof J Egbert as Amici in Support of Neither Party, *In re Trump*, 928 F3d 360 (4th Cir 2019). The Sixth Circuit also took the unusual step of requesting the parties to submit supplemental briefings presenting corpus linguistic analysis on the original meaning of Article III's Cases or Controversies requirement, see *Wright v Spaulding*, 939 F3d 695, 107 n 1 (6th Cir 2019).

meaning for judges in their interpretive tasks. They use the canonical no-vehicles-in-the-park rule from the Hart-Fuller debate (Hart 1958; Fuller 1958)[11] to demonstrate how the use of corpus linguistics can inform us of the ordinary meanings associated in this case with 'vehicle' and how this rule should be interpreted in relation to in/exclusion of what kind of 'vehicles'. The use of corpus-based data shows the size of the continuum of meanings that revolve around the notion of 'carrier' in the sense of transportation. Based on natural language corpora, a corpus linguistics study will also indicate which meanings are most likely to be relevant in the context of excluding vehicles from a park, presumably for safety reasons.

## 3.2 Semantic objectivity

A pragmatic approach to language is also essential in considerations of semantic objectivity. Legal language plays an essential part in the legal traditions based on the principle of the rule of law, namely that a legal system's scope is neutral and generally inclusive (Laske 2022). The law is said to be justified if it is verifiably general, neutral and impartial; it must be perceived as rational and meaningful. To that extent it is governed by the concept of objectivity, the character of which lies in the ability to consider or represent facts, information, etc., without being influenced by subjective elements such as particular perspectives, value commitments, community bias or personal interests, feelings or opinions, to name just a few relevant factors. The discussion of objectivity in this context is less concerned with what Reiss & Sprenger called 'objectivity as faithfulness to facts' (Reiss/Sprenger 2014). The argument evolves around semantic objectivity as the result of convention: the use of language and specific terms is deemed objective when it corresponds to the convention determined by the uses given by most speakers of a given group. The objectivity of the use of language and specific terms in a given situation can in turn be evaluated in relation to its conventional use. However, semantic objectivity in this sense does not mean that this conventional use of language and meaning is intrinsically free from bias. Language cannot be entirely neutral but is always the result of textualizing meaning in particular contexts. If we accept semantic objectivity as an underlying premise in law, in the case of written legal texts whether in the form of legislation, statutes or case law, the

---

11 The controversy was over legal positivism as argued by Hart, namely, to separate law and morality, and natural law as defended by Lon Fuller, that the law's binding power lies in morality. The no-vehicles-in-the-park thought experiment reveals interesting questions about the nature of legal language and, by implication, legal interpretation and legal rules.

legal language reveals, among others, how this objectivity is encoded and decoded. Studying the legal language and how meanings are formed in that particular context enables us, inter alia, to ascertain aspects such as intersubjective agreement of the use of language, as well as partiality. There is a need to 'measure' such aspects empirically, rather than just according to our intuition. Text and terminological analysis using big data methodologies, such as electronically held corpora and linguistics concordance software, allows us to gain better empirical insight into how meanings are encoded, while simultaneously supplying textual context. If combined with a diachronic approach, we will also learn of historical usages from periods of which we have little linguistic experience. This highlights how semantic objectivity in law can be relative, influenced by time and place, but it will also provide the contextual information that can help us evaluate objectivity at a given moment in relation to its own time and place. The corpus linguistics approach does not preclude other approaches such as, for example, discourse analysis. On the contrary, the two could be used as complementary tools, which is particularly relevant in assessing partiality and biases. Humpty Dumpty in Alice's Wonderland admits that he can chose language to mean whatever he wants it to mean. Recently, we have been confronted with the realities of how these sorts of self-appointed meanings have become prevalent. Corpus-based methodologies and tools provide data that can reveal semantic prosody and manipulation of meanings, which, in turn, inform us on the presence or absence of elements such as semantic objectivity and biases.

## 3.3 Textual representation

One area where sociolinguistics is particularly relevant is in the textual representation in legislation of specific categories of persons and their right/legal capacity. As social governance is one of the fundamental roles of law, the theory is that the analysis of the language in which legislation is written will show how the vested interests are encoded in the source texts. To reveal these sociolinguistic aspects of legal texts it is essential that linguistic research on textual representations is a part of analysing the sources beyond content and the purely descriptive.

The normative framework that regulates a society is informed by the prevalent needs and social attitudes, usually of those who hold the socio-political and economic power. Law is, therefore, not just social governance, but also social engineering, frequently at the service of the powerful and to the detriment of underprivileged groups of society. Traditional legal scholarship shows indications of whose rights and legal capacities are protected and exposes those which are not. From a social engineering point of view, this displays the level of importance attributed to each element; the more importance is given to a particular right/legal capacity, the more protection will be afforded to it. But beyond the prescriptive and normative content of laws and their application through the judiciary, a fundamental ingredient of the rights and constraints experienced by particular groups of people is the social perception of their identity as a group. It is within that context that evaluating the language and discourse used in laws and their application will reveal how particular categories of people are spoken about and the reality of their rights and constraints based on the social perception as individuals and as a group. This approach places sociolinguistic aspects of language and pragmatics at the centre of legal research. The underlying conceptual approach lies in textual representation providing new insights from a completely different angle than traditionally practiced in legal scholarship, namely that of decoding meanings and underlying discourses embedded in the law.

The situation of women's legal status and legal capacity is a case in hand. Adopting textual representation as a conceptual approach when researching the capacities afforded or denied to women, in terms of the normative/prescriptive (e.g. customs, laws, rules) and in the exercise of legal authority (e.g. making contracts), justice (e.g. access to law courts) or non-judicial petitions. When legal capacities are encoded in laws/rules, the constraints on women are normative and the experience lies in the confrontation with formal laws/rules. When the constraints are contained in the application of the law or the judicial process, the experience is one of law in action. However, the constraints experienced due to biased attitudes against women both as individuals and as a group depend on the discourse with which the laws, legal processes and judicial procedures are linguistically expressed and applied. The nature of this third kind of constraint lies in the language of the source texts that are drawn up by men and in an entirely male socio-professional and institutional context.

Studies on the textual representation of medieval women in (customary) laws have revealed some interesting aspects of the legal capacity[12] women enjoyed or lacked in thirteenth–century Normandy, Saxony and Norway. In the case of *Le Très Ancien Coutumier de Normandie*, it was not only interesting to obtain the data for the representation of the female 'voice', but also useful to compare that data with the data collected for the equivalent male descriptors. In fact, it was that very comparison that also drove home just how minimal and passive the female representation is in the Coutumier (Laske 2020,[b] 29–54). Using corpus linguistics methodologies, the text was searched more specifically for terms that are

---

[12] Legal capacity is defined as the legal right to make particular decisions, which are usually linked to responsibilities of a legal nature.

likely linguistic expressions of women's (legal) positions or would reveal such positions. In the case of some keywords, these results were compared to equivalent terms for men. The descriptors for women can roughly be classified into three main groups moving from the central to the peripheral. At the centre, the terms and language relate to:
(a) the woman as her own person: e.g. she, woman.
It then moves to:
(b) the definition of herself in relation to others: e.g. sister, daughter, mother, eldest, cousin.
On the more peripheral level, we find language that relates to her:
(c) legal/marital/sexual status: e.g. widow, virgin.

Also included in this category are terms that are commonly found in the context of women's property rights, such as dower and inheritance.

From the word list produced by the AntConc software based on the Coutumier corpus compiled from the source referenced in the bibliography below, we can see that there are 10 categories of descriptions of women from among a total of 215 occurrences. This includes terms such as 'she', 'woman', 'sister', 'daughter', 'mother', 'widow', 'girl', 'eldest', 'virgin', and 'cousin'.[13] In contrast, there are 21 categories of descriptors[14] for men, totalling 1,154 occurrences for a greater variety of terms than used to describe women.

From these figures alone, we can observe that the male voice predominates in comparison to the one relating to women. The pronoun 'il' is ranked second on the word list with 601 hits. Although on occasions, these may not relate to a male descriptor but rather to the neutral scenario of, for example, 'it has been said' ("Il a esté dit devant generalment qe . . ."Chapitre LVII ), the occurrence of this pronoun (331 per 10,000 words) occurs approximately eight times as frequently as its female counterpart (44 per 10,000 words), which is ranked 39th with 70 hits and its plural form – 'elles' – with 10 hits on rank 244. The contrast is less striking for the generic descriptor woman – man. We have 63 hits for 'fame/fames', which represents 35 per 10,000 words, and 107 hits for 'homme'[15] in its various spellings, which is 59 occurrences per 10,000 words.

---

13 The terms are: elle/elles; fame/fames; suer/suers; fille/filles; mere/mère; mechine/ meschine; vueve/vueves; ainznee; pucelle; cousine.
14 The terms are: il; homes/home/homs/hom/hommes/homme/omme; duc; roi; segneur/segneurs/seigneur; chevalier/chevalier; evesque; sire/sires; freres/frer/frères/; peres/pere/ père; filz/fill/fil; mari/mariz; senechau/senechal; sergent; baron/barons; arcevesque/ arcevesques; ainznez/ainzné; borjoir; prestres; cousins/cosins/cousin; neveu.
15 homes/home/homs/hom/hommes/homme/omme.

However, it is interesting to observe that there is also a greater variety of categories of descriptors for men than for women. In other words, not only is there a considerably greater presence of the male voice in the text, but it also reveals that men have a greater variety of roles. The references to women are always in their relation to the family group or their marital status: women are sisters, mothers, daughters, cousins, widows, virgins, etc, 10 categories in all.[16] Their social status is revealed through their family and marital status, never through descriptors saying anything in relation to other kinds of status, such as power, possession, economic activity. Terms that express some kind of female lordship or hierarchy, such as 'dame', 'domina', are entirely absent from the text. Men, on the other hand, are identified in relation to several roles (21 categories).[17] Besides the family relation descriptions such as 'father', 'brother', 'son', 'husband' (7 categories), men are listed in relation to socio-political and religious power, such as 'king', 'earl' or 'knight', '(arch)bishop', 'priest', 'seneschal'. In other words, the narrative told of men in *Le Très Ancien Coutumier de Normandie,* is not only more substantial but also considerably more varied and thus contextually richer.

This is also borne out by the data shown in Figure 1, which details some of the gender descriptors.[18] We can observe that the term 'fame' not only occurs considerably less than its male equivalent but also approximately as often as 'duc' and only slightly more than 'roi'. In other words, a generic term that describes approximately half the population occurs roughly as often as the term 'duke', which is used to describe only one person/office in Normandy. The family-related descriptors are also dominant for men: 'fils', 'pere', 'frere', 'mari' are all ahead of their female equivalent, except for 'suer', which occurs almost as often as 'frere'.

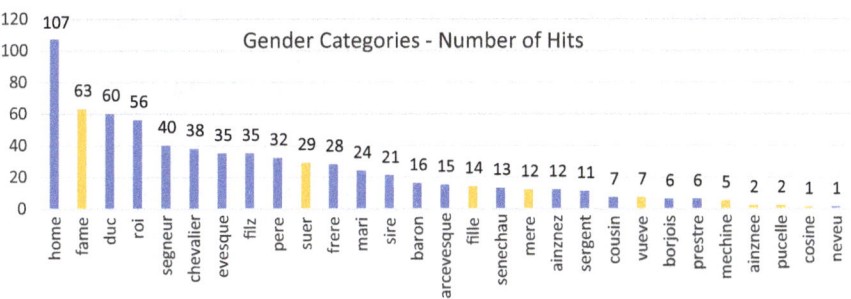

**Figure 1:** number of occurrences for all gender categories in *Le très ancien coutumier de Normandie.*

---

**16** see n.26.
**17** see n.27.
**18** All spelling variations were included.

Despite the considerably higher textual presence of references to men, we can observe from the proximity searches that certain terms referring to matters of a legal nature can be found far more often in contexts relating to women than men. While we may expect to find references to women mentioned more often in discussion relating to 'dower' ('doere') or the 'third' ('tierz') of her husband's estate that she can expect will be settled on her, it is surprising that the terms 'heritage' and 'mariage' are also predominately associated with women. One would expect men to be involved as much as women in matters of inheritance and marriage, but the proximity figures do not appear to bear that out. It would, therefore, be reasonable to conclude that discussions of marriage, inheritance, dower, and the third settled on married women mainly relate to women. A closer look at the textual context in which the terms 'fame' and 'home' occur confirms the findings of the proximity study. In over half of the KWIC lines for women, the context relates to matters of inheritance, family, dower, or marriage. However, issues of land and property, although at the heart of the discussion about inheritance and dower, are more associated with men. So are court and legal proceedings, crimes, and violence, as well as matters of political power, feudal homage, and duties – the last category is entirely absent in the KWIC lines for fame. A detailed analysis of the KWIC lines also reveals the types of processes with which the search terms associate. Without going into the specific statistics, it can be said that the female descriptors tend to be governed by verbal groups that reflect a more passive rather than an active aspect of the experiential world. Women are depicted in passive roles: things are being done to them, or others act on their behalf, e.g. "elle a esté prise a force" ("she was taken by force"), "elle doit estre envoiee a l'evesque" ("she must be sent to the bishop"). In this context, it is also interesting to examine the context for adjectives, as they tend to express attitude and opinions. The Norman Coutumier did not provide any significant data on that point, but it can be relevant in the case court reports or sources that have a more narrative element.

The female voice is rather feeble in *Le Très Ancien Coutumier de Normandie*, which is particularly striking in the comparative data sets that highlight the extent of the discrepancies when set side by side with the equivalent male terms. Although in the context of some matters, references to women are considerably higher than to men, the study of the processes that dominate the relevant KWIC lines, the verbal groups tend not to reflect active participation. A priori, this linguistic study of the Coutumier confirms the assumption that women in thirteenth-century Normandy had relatively low legal capacity and found themselves marginalised and in legal/social dependencies on men, especially husbands.

Vested interests encoded in the textual representation of women can be observed in the thirteenth-century Saxon customary law text *Sachsenspiegel*.[19] In general, these late medieval customary law texts form part of the hegemonic divisions that govern the relevant societies. The procreation imperative is the basis for marriage and the distinction between legitimate and illegitimate children for the purposes of inheritance (Caviness/Nelson 2018, 140). This in turn is a key element to territorial controls, ownership within families, and the feudal set-up of the land-owning classes and their vassals (Caviness/Nelson 2018, 172). Binary and heterosexual constructs are concrete historical realities entrenched in the "patriarchal kinship systems, by both civil and religious law, and by strenuously enforcing deeply entrenched values and taboos" (Fryre 1993, 493).

In the Rhymed Preface of the *Sachsenspiegel* a reference to the book's title reflects, so to speak, the binary gender divisions:

> "Spegel der Sassen"
> Scal dit buk sin genant,
> went Sassen recht is hir an bekant,
> Alse an eneme spegele de vrowen
> er antlite scowen.[20]

The symbolism of the mirror has different associations for men than it has for women. The law book is described as a mirror that reflects the familiar and in which Saxons see their laws and customs, with which they identify as Saxons. For a woman, the mirror reflects just her own image, which arguably refers to her vanity rather than to more high-minded matters such as Saxon laws. It also emphasises the fact that the law book is addressed to men; it may deal with women at times, but they were not really in a position to speak for themselves and doing so could have long-term consequences (see Califurnia discussed below). In this way, the binary opposition is compounded with male dominance inherent in the hegemonic structure. To that extent, the *Sachsenspiegel* and other customary law books codify and maintain this 'otherness' and the unequal rights inherent in its construction. The text is of additional interest for the concept of textual representation to the extent that there are four extant picture manuscripts with images representing the underlying discourse, attitudes, biases and values held by the

---

**19** The *Sachsenspiegel* (original title: *Sassen Speyghel* or *Sassenspegel*), attributed to Eike von Repgow (c. 1180–1235), was compiled after 1220 in eastern Saxony. Whether he was the sole author is still a moot point. It comprises the regional customary laws, the so-called *Landrecht* often divided into books I–III, and the feudal laws affecting the gentry, the so-called *Lehnrecht*, sometimes cited as book IV.
**20** "'The Mirror of the Saxons' / Shall this book be called / because with it the Saxon laws will come to be known / like a mirror in which women can see their countenance."

male landed classes. These manuscripts were all produced later[21] than Eike's work and can therefore not be attributed to his direct influence; indeed, we have no evidence of what any picture manuscript contemporary with Eike might have looked like. But the fact that from 70 to 150 years later such richly decorated manuscripts were still being commissioned and produced is a good indication that the text remained authoritative. The picture manuscripts also reveal that attitudes towards women hardened during the period between the original text and the subsequent picture manuscripts. The pictures in these manuscripts were juxtaposed and arranged in a way that gave similar prominence to the text and the pictures. Manuwald (2005) had coined the phrase 'zweisprachige Ausgabe'[22] rather than as an illustrated book and that applies very appropriately to the four extant picture books. The book's ideological content was driven home by the written textual representation as well as by the intervisuality with a visual discourse, the pictures being "in dialogue with the text rather than as literal illustrations of it and capable of inflecting as well as reflecting its reception in the time and place of their iteration" (Caviness/Nelson 2018, 43). In the pictures, legal meanings were encoded in gestures and body language (von Amira 1909) and if we follow the premise of the 'zweisprachige Ausgabe' we need to consider text and picture as part of a whole textual representation.

A particularly interesting example concerns the ancestral transgression of Calefurnia who supposedly forfeited that right for generations of women to come: instead of addressing the court through her guardian, she speaks to the emperor directly and furiously, thus providing justification for denying women a direct voice in a court of law. Of particular interest is how the visual discourse mirrored the text, as can be observed from the Heidelberg manuscript (Figure 2). The medieval conception of women as sedentary and typically associated with phlegm, the cold and moist humour, predisposed them to become angry slowly but violently (Cadden 2003, 183–186). In that case, guardianship was necessary to control them rather than to protect them (Westphal 2002, 106). This was encoded in the pictorial representation of Calefurnia. According to Karl von Amira's study of hand (arm) gestures in the *Sachsenspiegel*, Calefurnia's finger points (von Amira 1909, 264 gesture 5a and b) at the emperor: the dispute is *with* him rather than a referral *before* him. The emperor's hand also points at Calefurnia: this indicates her disrespectful behaviour, an element further emphasised by a kind of

---

**21** These manuscripts are commonly named after the towns where they are preserved: Heidelberg (H) 1294–1304, Dresden (D) 1295–1363, Oldenburg (O) 1336, Wolfenbüttel (W) 1348–62/71.
**22** Literal translation: 'bilingual edition', or alternatively it could be translated as 'bilingual recension'.

**Figure 2:** *Sachsenspiegel Landrecht*, Heidelberg manuscript, 10v.
https://doi.org/10.11588/diglit.85#0034 (public domain)

brush protruding from her behind, as though she were exposing her private parts to him (von Amira 1909, 264 gesture 5b). This implication was recounted in some versions of the *Schwabenspiegel*, compiled half a century after the *Sachsenspiegel*, but about two decades before the Heidelberg manuscript (Koschorreck 1976, 80). Here, intertextuality and intervisuality embody a powerful message. A woman audacious enough to address the emperor directly, without representation by her guardian, had to bear the consequences of being excluded from a court of law, not only for herself but also for generations of women to follow. Calefurnia's act was not only contrary to customary law but further sullied by reference to her private parts. The textual provision did not invoke this circumstance, which was added later in the pictorial representation.

A further intervisual element of Calefurnia's protruding brush lies in its similarity to an animal tail. In the text, the passage about Calefurnia was preceded by seemingly unrelated provisions on keeping and controlling animals, such as dogs, wolves, deer, bears or monkeys, and on damages for which a keeper was liable if such animals caused any harm.[23] In the scan from the Heidelberg manuscript (Figure 2), the pictorial narrative can be clearly followed from the dangerous animals mentioned in the text attacking a person (at the top of the page) through to Calefurnia with her bushy tail. A further association is made with the scruffy rape victim at the very bottom of the pictures, related to the rape provision of *Sachsenspiegel Landrecht* II, 64, 1. It is interesting to note that the pictorial representations deal more harshly with women than the initial text. This indicates a hardening discourse on women during the century that followed the writing of the *Sachsenspiegel*.

Systemic inequalities by virtue of factors such as birth, of belonging to a particular estate, of gender or category of persons were perpetuated by the *Sachsenspiegel*'s legal text. But experience of this social asymmetry was further generated and controlled by the generalised and dominant discourse that explicitly taught inequality and reminded people of their station in life. This was embedded in the visual discourse by juxtaposing dangerous animals in need of control and Calefurnia's transgression with what looks like a bushy tail. It exemplifies how the dominant discourse was forged by the control of women's action – they could not act without a guardian – and the control of the social perception of them as a group – transgression of the guardian principle had grave long–term consequences, and women represented in action were visually linked to dangerous animals that needed to be controlled.

---

23 *Sachsenspiegel Landrecht* II, 62.

By concentrating on specific texts, the two studies discussed so far were akin to a micro rather than a macro textual approach. While the (socio) linguistic angle provided context to specific provisions and concepts within each text, it is of interest to find further context through comparisons with similar or contrasting texts, both synchronically or diachronically. During the same thirteenth century but further to the north King Magnus Hakonarson of Norway (ruled 1261–1280) adopted a new national law book for the entire kingdom, the *Lanslög* (laws of the land). While this is a top down piece of royal legislation rather than a text fixing in writing usages that a community had accepted over time, the linguistic study of gender markers in the texts have revealed some interesting differences. In contrast to the Norman and Saxon texts, the Norwegian legislation shows a high proportion of gender-neutral language.[24] This can be observed when examining, for example, the word 'maðr' which translates as 'someone' or as 'man' including the meaning 'mankind'. In other words, 'maðr' has both a gender-neutral meaning, as well as a specifically masculine one. The source text contains 45,445 word tokens[25] and the word 'maðr' appears 404 times. A detailed contextual analysis to determine more precisely the gender encoded in the reference showed that nearly 70% of the references (282 out of 404) were deemed to be gender-neutral, as shown in Figure 3 below. For the purpose of that study, the concordance lines were grouped into four categories of meanings, although in some instances it was impossible to determine the gender marker:

– gender-neutral meaning,
– use of 'maðr' probably referring to men,
– use of 'maðr' referring specifically to men,
– use of 'maðr' probably referring to women.

No instances could be found where 'maðr' referred exclusively to women, although there are very few meanings of 'maðr' likely to include women. It would be reasonable to conclude that, as in the contemporaneous Norman and Saxon texts, women are underrepresented in this legislation, if it were not for a large number of gender-neutral references that may well include women. The frequency data is interesting but in itself insufficiently definite without further contextual information. A study of the textual context of the 404 instances of 'maðr' has confirmed that its use in the gender-neutral sense of 'person' outstrips the instances where gender is marked (see Figure 3).

---

[24] The detailed results of this study will be published in a forthcoming paper.
[25] Word tokens represents the number of individual words in a text.

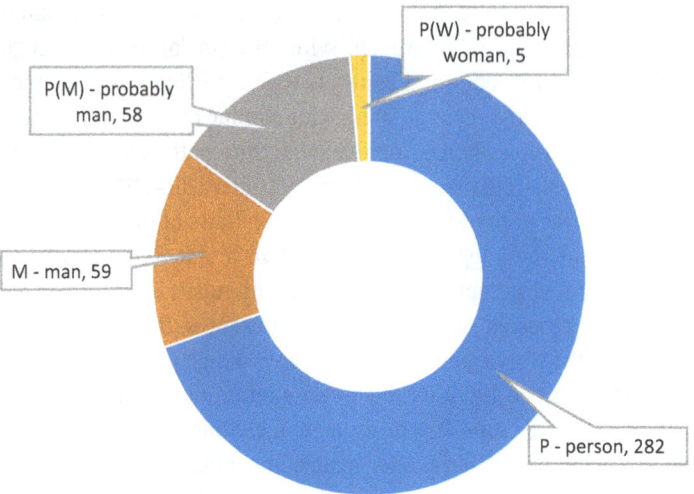

**Figure 3:** Contextual meanings of the term 'maðr'.

The statistical distribution of the occurrence of 'maðr' throughout the various chapters in the text can be found in Table 1. As the chapters are of different lengths, the number of hits for 'maðr' must be contextualised by the relative frequency for each chapter, which is calculated on the statistical basis of the word occurring per 10,000 words and is shown in the last column of the table.

The data presented in Figure 4 offers information on the distribution throughout each chapter of the gender markers that can be attributed to 'maðr' by examining the textual context in which the term occurs. The green column represent the number of hits per chapter. The occurrences of 'maðr' have been grouped into the four categories of meanings described above. The proportionality between the number of hits for 'maðr' (in green) and the various categories of meaning, in particular the gender neutral one is of particular interest to observe in this figure.

A study of the way the term 'maðr' is distributed throughout the text shows that the highest relative frequency can be found in section 10 on thieves (Þjófabǫlkr; 208 per 10,000 words). While criminal law and provisions tend to apply to both sexes, although sometimes the punishments may differ, it is striking in this text that the use of 'maðr' with male gendered markers outnumbers the gender-neutral use. It is the only chapter where this can be observed. Yet, the text is considerably less gendered in its criminal law provisions in comparison to other contemporaneous texts.

**Table 1:** Hits for maðr' throughout the Lanslög.

| Chapter | English translation | Total words | 'maðr' (hits in chapter) | 'maðr' (relative frequency in chapter, per 10,000 words) |
| --- | --- | --- | --- | --- |
| Entire text | | 45,445 | 404 | 89 |
| 1. Prologus | Prologue | 673 | 1 | 15 |
| 2. Þingfararbǫlkr | Section about Attending Assemblies | 2288 | 9 | 39 |
| 3. Kristinsdómsbǫlkr | Christianity Section | 2649 | 0 | 0 |
| 4. Útfararbǫlkr | Land Defence Section | 4106 | 33 | 80 |
| 5. Mannhelgi | Human Inviolability | 6691 | 85 | 127 |
| 6. Erfðatal | The List of Inheritance | 5671 | 40 | 71 |
| 7. Landabrigði | Land Redemption | 3113 | 21 | 67 |
| 8. Landsleigubǫlkr | Land Tenancy Section | 12762 | 124 | 97 |
| 9. Kaupabǫlkr | Trade Section | 4801 | 48 | 100 |
| 10. Þjófabǫlkr | Thieves' Section | 1825 | 38 | 208 |
| 11. Um réttarbǿtr konunga | Law Amendments | 634 | 5 | 79 |
| 12. Capitulus | Epilogue | 232 | 0 | 0 |

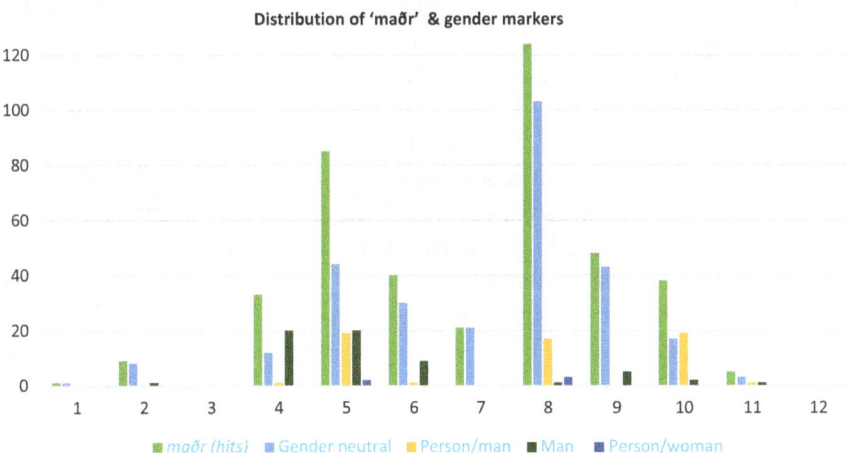

**Figure 4:** Distribution of maðr & gender markers throughout the chapters of the Lanslög.

For example, in the *Sachsenspiegel* the definition of a red-handed crime only refers to men:

"De hanthafte dat is dar, swar men enen man mit der dat begript oder in der vlucht der dat . . ."[26]

Frequently provisions specifically refer only to men (in blue):

"Sleit en man den anderen dot dorch not, unde ne darn he nicht bi em bliven . . ."[27]

Many provisions typically start with 'those who . . .' or 'one who . . .', but this seemingly neutral descriptor (in green) is rapidly supplanted by a male pronoun (in blue):

"Swe des nachtes korn stelet, de verscult des galgen; stelet he it des dages, it geit eme an den hals."[28]

Other provisions may, in theory, be interpreted more gender neutrally, but in view of the textual context and the fact that when women are included, this is clearly signposted, it is reasonable to conclude that the provisions only apply to men, unless it includes specifically women (in red):

"De den man sleit oder vet oder rovet oder bernet sunder mordbrant, oder wif oder maget nodeget, undebrekere, unde de in overhure begrepen werdet, den scal men dat hovet af slan."[29] The words in green could be interpretated as gender-neutral, but it is not unreasonable to conclude that in view of the context and violence of the crimes described, the provisions were written with male rather than female perpetrators in mind.

Of particular interest is the fact that the trade section (Kaupabǫlkr [9]) has a high relative frequency of hits for 'maðr' and that a great majority of these occurrences can be categorised as gender-neutral. Legal capacities to incur debt, hold assets, conclude contracts, etc. were not granted to women in many other areas of Europe, usually because women, in particular wives, were subject to guardianship, first of their fathers/brothers and subsequently of their husband. For example, in the *Sachsenspiegel* an all-pervasive guardianship limits the legal capacity

---

[26] *Sachsenspiegel Landrecht* II, 35: "A violation is red-handed when someone intercepts a man in the act of a violation, or in the flight from the deed . . .".
[27] *Sachsenspiegel Landrecht* II, 14§1: "If a man slays another in self-defence and cannot remain with him . . .".
[28] *Sachsenspiegel Landrecht* II, 39§1: "Anyone who steals grain in the night earns the gallows; if he steals it during the day, he pays with his neck."
[29] *Sachsenspiegel Landrecht* II, 13§5: "One who beats or abducts a man, or robs, or commits arson (with the exception of murderous arson), or rapes a woman or girl and violates the peace shall be beheaded."

of women to control their property[30] or take an active part in the judicial process.[31] Even when women have a certain lee way to act, such as girls and unmarried women who could sell land they owned without the permission of their guardian, the blanket restriction to act in a court of law meant that they could not fully exercise that legal capacity.[32] It is therefore surprising that in the Norwegian law, the gender-neutral language seems to indicate that women were not excluded from legal capacity and economic activities.

A similar situation can be observed in relation to the sections on the list of inheritance, land redemption and land tenancy Erfðatal [6], Landabrigði [7], Landsleigubǫlkr [8]). In these sections the blue column representing the gender-neutral meanings is of the same or similar height to the total number of hits for 'maðr' (green column). It is an indication that holding (allodial) land as well as inheritance and land tenancy provisions potentially addressed women, though not directly nor exclusively. This is in stark contrast to the *Sachsenspiegel* where on marriage, a man takes into his possession all his wife's property in lawful guardianship (*Sachsenspiegel, Landrecht* I, 31§2).

The data relating to land defence (Útfararbǫlkr [6]) shows a high percentage of male gender markers in the use of 'maðr'. This, in itself, is not surprising as defence has traditionally been the men's domain. No women were involved in matters relating to army, warfare and land defence. Yet, as we can observe from Figure 4, the use of 'maðr' has a neutral gender marker in a third of the hits in that section. While women were excluded from matters of war, the fact that they could hold and work land meant that they were liable to contribute to taxes and the efforts of shipbuilding and warfare in general. This is a brief outline of research in progress which will be expanded to other terms and gender markers. The initial study on 'maðr' provides interesting data to observe a trend of gender neutrality that cannot be found in other contemporaneous Northern European texts.

---

**30** *Sachsenspiegel, Landrecht* I, 45§2 stipulates that a wife can not alienate or sell her property/ land, nor transfer land held in life estate without her husband's consent.
**31** *Sachsenspiegel, Landrecht* I, 46 prescribes that girls and married women must have guardians to bring a suit because one cannot prosecuted them with witnesses for what they say or do in a courts of law.
**32** *Sachsenspiegel, Landrecht* I, 45§2.

## 4 Conclusions

The discussion in this paper has highlighted how a more sociolinguistic approach to legal linguistics contributes insights into the law that have hitherto been under-explored, in particular, in legal scholarship. Considering legal language as language in use rather than only as one used for special purposes, underlines the fact that it is also constituted of underlying meanings and sociolinguistic content that can be revealed through the use of big data analysis and corpus linguistics.

The studies described in the third section show the potential of using a sociolinguistic lens for understanding the law. The application of corpus linguistics in the search for ordinary meanings in statutory interpretation and judicial decisions implies a move away from specialised language and towards the search for meanings in everyday language use. In the case of semantic objectivity, the use of language and specific terms is deemed objective when it corresponds to the convention determined by the uses given by most speakers of a given group. In this sense, the convention appeals to language in use, including the underlying meanings and discourse encoded in the convention. This can be examined by using corpus linguistics methodologies. The research on textual representation as a conceptual tool to understand the de facto experience of law, going beyond the content analysis of legal texts, is work in progress. It is a large scale study on the textual representation of women's legal capacity in legal documents and hypothesises that the reality of women's legal capacity lies in the experience of the attitudes/biases they face, imposing constraints on their capacity to act with legal authority, for example, in relation to their income-generating activities. The legal capacity within which women can de facto operate is conditioned by the constraints encoded in the rules and laws of the prescriptive/normative texts. In the case of constraints within the judicial process and procedures, their impact become apparent when engaged in these procedures. However, constraints experienced due to biased attitudes against women as a group depend on the discourse with which the laws, legal processes and judicial procedures are linguistically expressed and applied. The nature of this third kind of constraint lies in the sociolinguistic aspects of the language of the source texts, which are drawn up in a male dominated socio-professional and institutional context. The sociolinguistic study of the legal language will show how the vested interests are encoded in the language and in the textual representation of the source texts. While this approach may be common in social sciences, in legal research it is still under explored, yet it would provide valuable contextual information on the understanding of law, its function and its workings.

# Bibliography

Baaij, Jaap (ed.). 2012. *The Role of Legal Translation in Legal Harmonisation*, Alphen aan den Rijn, Kluwer.
Bhati, Vijay, Christoph A. Hafner, Lindsay Miller & Anne Wagner (eds.). 2012. *Transparency, Power and Control: Perspectives on Legal Communication*, London, Routledge.
Biel, Lucja & Agnieszka Doczekalska. 2020. *How do supranational terms transfer into national legal systems. A corpus-informed study of EU English terminology in consumer protection directives and UK, Irish and Maltese transposing acts*, Terminology 26(2), 184–212.
Cadden, Joan. 2003. *Meanings of Sex Difference in the Middle Ages. Medicine, Science and Culture*, Cambridge, CUP.
Caviness, Madeline & Charles Nelson. 2018. *Women and Jews in the Sachsenspiegel Picture-Books*, Turnhout, Brepols.
Coulthard, Malcom Alison Johnson (eds.). 2012. *Routledge Handbook of Forensic Linguistics: Language of Evidence*, London, Routledge.
*Coutumiers de Normandie*, Ernest-Joseph Tardif (ed.). 1903. 2 tomes, 3 volumes; tome I, $2^e$ partie, *Le Très Ancien Coutumier de Normandie*, texte français et normand, Rouen/Paris, Lestringant/Picard, 1903.
Denning, Lord. 1979. *The Discipline of Law*, London, Butterworth.
Firth, John. 1957. *A Synopsis of Linguistics Theory 1930–1955*, Studies in Linguistic Analysis (Special Volume of the Philological Society), 1–32.
Freeman, Michel Fiona Smith (eds.). 2013. *Law and Language*, Oxford, OUP.
Frye, Marilyn. 1993. *Virgin Women*. In Alison Jaggar & Paula Rothenberg (eds.), *Feminist Frameworks: Alternative Theoretical Accounts of the Relations between Women and Men*, New York, McGraw-Hill.
Fuller, Lon. 1958. *Positivism and Fidelity to Law – A Reply to Professor Hart*, Harvard Law Review 71, 630.
Giannoni, Davide & Celina Frade (eds.). 2010. *Researching Language and the Law: Textual Features and Translation Issues*, Berlin, Peter Lang.
Hart, Herbert. 1958. *Positivism and the Separation of Law and Morals*, Harvard Law Review 71, 593.
Halliday, Michael. 1976. *System and function in language*, London, Edward Arnold.
Halliday, Michael. 1985. *An Introduction to Functional Grammar*, London, Edward Arnold.
Halliday, Michael & Ruqaiya Hasan. 1985. *Language, context and text: Aspects of language in a social semiotic perspective*, Geelong, Deakin University Press.
Koschorreck, Walter. 1976. *Der Sachsenspiegel in Bildern aus der Heidelberger Bilderhandschrift: Ausgewählt und Erläutert von Walter Koschorreck*, Frankfurt a.M., Insel Verlag.
Laske, Caroline. 2016. *Losing Touch with the Common Tongues – the story of law French*, International Journal of Legal Discourse 1(1), 169–192, DOI 10.1515/ijld-2016-0002.
Laske, Caroline. 2018. *Le Law French, un idiome protégeant les privilèges du monde des juristes anglais entre 1250 et 1731*, Corela: Cognition, Representation, Language, 1–20.
Laske, Caroline. 2020[a]. *Law, Language and Change. A Diachronic Semantic Analysis of* Consideration *in the Common Law*, Leiden, Brill.
Laske, Caroline. 2020[b]. *Medieval Women in the Très Ancien Coutumier de Normandie. Textual Representation of Asymmetrical Dependencies*, Berlin, EB-Verlag.
Laske, Caroline. 2022. *Big data linguistic analysis of legal texts – objectivity debunked?*. In Gonzalo Villa-Rosas & Jorge Luis Fabra-Zamora (eds.), *Objectivity in Jurisprudence, Legal Interpretation and Practical Reasoning*, Cheltenham, Edward Elgar Publishing, 167–192.
Lee, Thomas & Stephen Mouritsen. 2018. *Judging Ordinary Meaning*, The Yale Journal 127, 788–879.

Lee, Thomas & Stephen Mouritsen. 2021. *The Corpus and the Critics*, University of Chicago Law Review 88(2).
*Landslǫg Magnúss Hákonarsonar*, Holm perg 34 4to, https://clarino.uib.no/menota/document-element?sessionid=254632958818230&cpos=2107889&corpus=menota
Manuwald, Henrike. 2005. *Die Grosse Bilderhandschrift des Willehalm: Kommentierter Text oder "zweisprachige Ausgabe"?*. In Britta Bussmann, Albrecht Hausmann, Annelie Kraft *et al.* (eds.), Übertragungen: Formen und Konzepte von Reproduktion in Mittelalter und früher Neuzeit, Berlin, De Gruyter, 377–394.
Mattila, Heikki. 2013. *Comparative Legal Linguistics. Language of Law, Latin and Modern Lingua Francas*, London, Routledge.
Mellinkoff, David. 1963. *The Language of the Law*, Boston, Little, Brown & Co.
Mori, Laura (ed.). 2018. *Observing Eurolects: Corpus Analysis of Linguistic Variation in EU Law*, Amsterdam, John Benjamin.
Ormrod, Mark. 2003. *The Use of English: Language, Law and Political Culture in Fourteenth-Century England*, Speculum 73, 3, 750–787.
Prieto Ramos, Fernando. 2013. *Legal Translation in Context: Professional Issues and Prospects*, Oxford, Peter Lang.
Recanati, François. 2004. *Literal Meaning*, Cambridge, Cambrideg University Press.
Reiss, Julian/ Sprenger Jan. 2014. *Scientific Objectivity*, The Stanford Encyclopedia of Philosophy, URL: https://plato.stanford.edu/entries/scientific-objectivity/ accessed 15. November 2022.
Rothwell, William. 1975. *The Role of French in Thirteenth-Century England*, Bulletin, John Rylands Library 58, 445–466.
Rothwell, William. 1993. *Language and government in medieval England*, Zeitschrift für französische Sprache und Literatur, 93, 258–270.
*Sachsenspiegel* (original title: *Sassen Speyghel* or *Sassenspegel*), attributed to Eike von Repgow (c. 1180–1235), https://digi.ub.uni-heidelberg.de/diglit/cpg164 or www.sachsenspiegel-online.de, also https://www.wdl.org/en/item/11620/. The translations come from Maria Dobozy's edition in English (1999) *The Saxon Mirror. A* Sachsenspiegel *of the Fourteenth Century*, Philadelphia: University of Pennsylvania Press.
Middle German printed edition, edited by K.A. Eckhardt (1933/1955) *Sachsenspiegel Landrecht*, Göttingen: Musterschmidt Verlag; and *Sachsenspiegel Lehnrecht*, Göttingen: Musterschmidt Verlag (1933/1956).
Solan, Lawrence. 2016. *Can Corpus Linguistics Help Make Originalism Scientific?*, Yale Law Journal Forum 126, 57.
Solan, Lawrence & Tammy Gales. 2017. *Corpus Linguistics as a Tool in Legal Interpretation*, BYUL Review, 2017, 6, 1311.
Tiersma, Peter & Laurence Solan (eds.). 2016. *Oxford Handbook of Language and Law*, Oxford, OUP.
Warr, John. 1650/1810, *The Harleian Miscellany*, London: Dutton (new ed., this pamphlet was originally from 1649, it contains the debate leading up to the 1650 Act).
von Amira, Karl. 1909. *Die Handgebärden in den Bilderhandschriften des Sachsenspiegels*, https://digi.ub.uni-heidelberg.de/diglit/amira1905.
Westphal, Sarah. 2002. *Bad Girls in the Middle Ages: Gender, Law, and German Literature*, Essays in Medieval Studies 19, no. 1, 106.
Woodbine, George. 1943. *The Language of English Law*, Speculum 18, 395–436.

# Section 3: Theories of sense and meaning for legal investigations

Weronika Dzięgielewska and Wojciech Rzepiński
# What is practical about law? Contemporary legal philosophy on legal practice

**Abstract:** The paper aims at addressing the problem of how the legal philosophers use the concept of practice when describing law, and why 'practice-oriented' outlook on law can bring about more complex understanding of the social dimension of law. Thus, the authors analyse the accounts of central figures in legal theory that employ the concept of legal practice (among others Hart, Dworkin, Raz or Pavlakos). More specifically, the research includes metatheoretical remarks on the vocabularies privileged by authors of the theories in scope, when it comes to usage of the notion of "legal practice". It is also observed that these theories remain unsatisfactory as to explaining who the subject of the legal practice is.

**Keywords:** legal practice, analytic pragmatism, legal philosophy, meaning-use relation, inferentialism

## 1 Introduction

At first glance, our title seems deceptive. There are many obvious ways in which the law can be, or purports to be, practical. Probably the most prominent answer to our question is that the law aims at guiding and coordinating the behaviour of its addresses on a highly institutionalised level. It may seem that not much more could be said about law's practical side. However, our focus is metatheoretical, rather than object-level — we are less concerned about the law itself than about why the concept of practice is so generously and eagerly applied by legal philosophers for the purposes of describing law. We believe that adopting a novel methodological tool to analyse how the concept of practice is used in legal scholarship can shed new light both on the metatheories of legal practice and the assumptions regarding the social dimension of law made by legal theorists.

---

**Note:** The chapter is based on a significantly modified paper entitled: "*In search of the basic unit of a legal practice*" presented by Wojciech Rzepiński during the 5th ILLA General Conference: Language and the Law in the Age of Migration held between 7th and 9th September 2021 in Alicante. It was prepared within the framework of a research project funded by the National Science Centre (Narodowe Centrum Nauki) (PRELUDIUM 17, 2019/33/N/HS5/01418).

---

**Weronika Dzięgielewska, Wojciech Rzepiński,** Adam Mickiewicz University Poznań, Poland

https://doi.org/10.1515/9783110799651-009

The appeal of the concept of practice in contemporary legal theory seems apparent. More than 20 years ago Brian Z. Tamanaha noted: "For some time the basic tool used to analyse law has been the institution. Now there are additional concepts – practices, and interpretive communities and their shared meaning systems – which help open up new dimensions and draw different lines. Practices add an *activity* related dimension that cuts across the institution at many different levels" (Tamanaha 1999, 174). This passage confirms the intuition that engaging with the "practice-oriented" outlook on law is directly intertwined with the theoretical commitments underpinning how the social dimension of law is understood.

A framework for a novel metatheoretical analysis of the theories applying the concept of practice can be found on the grounds of analytic pragmatism, which has its roots in the philosophy of language. In his *Between Saying and Doing*, Robert B. Brandom, the father of analytic pragmatism, proposed to extend the project of language analysis by drawing attention to the *use* of expressions. This led him, interestingly, to focus on the agents who are responsible for those uses. As a means of analysing the practice of using concepts, Brandom introduced *meaning-use relations* – a useful tool, which – in our view – can be applied as a method for analysing legal theories. Of course, we acknowledge the differences between the philosophy of language (for which this tool was created) and the analysis of claims in the field of legal philosophy. Therefore, employing meaning-use relations for our present purposes should take into account the specificity of legal philosophy.

Accordingly, in the first part of the paper, we describe the idea of meaning-use relations. Particular emphasis will be placed on how this original tool proposed by Brandom needs to be adjusted to suit the requirements of legal theory.

Second, we will apply meaning-use relations as a tool for analysing the *uses* of the concept of practice that can be identified in the theoretical accounts put forward by the central figures in contemporary legal theory. The aim is to broaden the general understanding of which meaning is ascribed to the concept of practice in these theories, and to show the universal applicability of the proposed tool.[1] In order to unpack the ascriptions of meaning by each author, we identify the individual uses of the concept in the selected theories. Due to the limited scope of this chapter, we have chosen only four examples of employing the concept of practice for building a theory of law. However, we selected the most notable, namely: Hart's practice the-

---

[1] We assume that the uses of words confer the content of concepts, insofar as the epistemic access to the concepts is guaranteed through the words (see e.g. Pelc 1982, 36). Therefore, we remain in the later Wittgensteinian tradition which treats the meaning of words as how they are used in language (see Wittgenstein 1953, §43). Ascribing meaning, from this perspective, is understood as an act of making an assertion about the use of a concept by another language-user and the content she is believed to confer by this use (for example, by using 'that' clauses).

ory of rules, Dworkin's interpretation of legal practice, Raz's account of practices as grounding legal rules and values, and George Pavlakos' Practice Theory of Law.[2] Analysing these theories through the prism of the Brandomian relations should provide answers to the following questions (in the light of the selected theories):
1) What one must *do* (in terms of *practices-or-abilities*) in order to count as *being* part of the selected practice, and what do these *practices-or-abilities* allow the practitioners to express?
2) What is the *vocabulary* which *specifies* those *practices-or-abilities*?
3) What are the methodological consequences of committing oneself to specific vocabularies when describing legal practice?

Third, we will summarise the benefits of an analysis of legal practices (and other legal concepts) via meaning-use relations. Our main assumption is that this tool draws attention to the social level of the practice, that is, to the actions and abilities of its participants. Apart from the insights as to the meaning of the concept of practice articulated in the selected theories, we aim to show the overall plausibility of the proposed method for analysing the uses of legal concepts made by legal theorists.

# 2 MURs as a metatheoretical tool

## 2.1 Analysing theories

In order to provide an analysis of the uses of the concept of practice made by the selected legal theorists, we decided to apply meaning-use relationships ('MURs'). MURs were introduced to extend the project of classical (language) analysis, which is perceived by Brandom as "mak[ing] of the meanings expressed by *one* kind of locution in terms of the meanings expressed by *another* kind of locution" (Brandom 2008, 1). Instead of this approach, Brandom suggests that we "(. . .) understand pragmatics as providing special resources for extending and expanding the analytical semantic project: extending it from exclusive concern with relations among meanings to encompass also relations between meanings and use"

---

[2] This choice may seem arbitrary. We try to mitigate this arbitrariness by choosing two prominent positivist and non-positivist theories, as representatives of their kind. Moreover, the presented analysis does not aim at an exhaustive presentation of the theories applying the notion of practice in the description of law. We are also aware of the whole tradition of analyses of the theories we have actually chosen, which we are unable to fully acknowledge in this place. Therefore we decided to set a very limited aim in this chapter, which is to show the advantages and universal applicability of the pragmatically mediated relations for analysing legal concepts.

(Brandom 2008, 8). Therefore MURs were thought of as 'bridges' between pragmatics and semantics, aimed at conceptualising in language what exactly happens at the level of linguistic acts. For Brandom, if we are to describe meaning in terms of use, the preferred path is through the relations between the vocabularies and abilities that are necessary or sufficient for using them. This is because in the inferentialist picture the vocabularies become contentful when they function in correct material inferences which are established in the process of using language by the participants of the practice.[3] The correctness of the application of a concept is judged by the other participants, who check whether a selected user makes correct material inferences with the applied concept. One can be perceived by others as a proper user of a concept if one can make correct material inferences – in other words, when one has the relevant skills sufficient or necessary to use this concept. Therefore MURs depict the relationships between vocabularies and practices which are relevant for the process through which concepts obtain content.

Before we proceed to describing the specific types of different MURs, we want to turn to the pressing question of how exactly MURs can function as a metatheoretical tool for analysing legal theories. For the purposes of providing an answer, we give three reasons why we think MURs are a viable analytic method of analysis. Moreover, we explain the three main assumptions that we adopt for implementing MURs to anaylyse the concept of practice in the selected legal theories.

The first reason is that MURs are, in the Brandomian picture, a method of performing a pragmatic analysis, which is more an analysis of uses than an analysis of meanings. This type of analysis, while focusing on uses, treats them as complex – that is, as compounded from other uses (or, in Brandom's terminology, practices-or-abilities; Brandom 2008, 33). Therefore it enables uses (in this case, the usage of a concept by a particular theory) to be analysed and treated as entering into relationships with other practices and vocabularies (e.g. as composed by other uses or reliant on specific vocabularies). This is because of the previously introduced assumption that the content of concepts is generated in the instances when vocabularies are used in inferential chains of reacting to those uses by different participants of the practice.

Secondly, inferentialist-pragmatist analysis is concerned with particular types of uses, which are mostly sayings. In other words, the uses that are the object of analysis are uniform with discursive practices-or-abilities, which are the "practices-or-abilities that count as deploying vocabularies, as conferring or applying mean-

---

[3] As Brandom points out, "The practices that confer propositional and other sorts of conceptual content implicitly contain norms concerning how it is correct to use expressions, under what circumstances it is appropriate to perform various speech acts, and what the appropriate consequences of such performances are" (Brandom 1994: xiii).

ings" (Brandom 2008, 33). Deploying vocabularies can therefore lead to the analysis of deploying concepts, because vocabularies are treated as expressing concepts (Brandom 2008, 46), insofar as they function in materially correct (meaning-conferring) inferences. In this sense, the vocabularies deployed on the grounds of the analysed theories can reveal the content of the concepts they assume, insofar as they are decoded in the practices-and-abilities which are necessary or sufficient for a user to be considered a skilful user of these concepts on the grounds of a theory.

Thirdly, the MUR analysis of theories draws attention to how uses of a certain vocabulary involve relying on other vocabularies and require exercising certain abilities (or participating in specified practices). If a theory is understood as composed of sets of vocabularies, it can be analysed as assuming the corresponding layer-cake structure of interrelated practices and vocabularies. The advantage of such an assumption is that the following analysis is more focused on how meanings are not solitary and how they depend on practices-or-abilities that construe their social dimension. As a consequence, this focuses attention on the participants involved in applying these vocabularies, which raises awareness of subjective influence on the content of discursive practices.

Taking the above observations a step further and applying MUR pragmatic analysis for the purposes of analysing the content of legal theories requires adopting the following methodological assumptions:

1) The theories under analysis are treated as languages composed of sets of vocabularies, and it is assumed that these vocabularies express the meaning of the concepts employed by these theories.
2) The analysis of the four chosen theories will concentrate on the use of the concept of *practice*. Since the pragmatic apparatus of MUR employs talk about the practices itself, in the remaining part of the text we will first address the analysed concept of practice on the grounds of the selected theory, and then apply MUR to explicate the practices or vocabularies which are interrelated with this concept.
3) The object level of the analysis is concerned with the legal theories themselves, therefore we will assume that the vocabularies, practices and abilities of a selected theory are interrelated. This reservation is necessary as these theories – which were created as a result of a certain social practice – address in their content a specific social practice, that is, legal practice. Accordingly, to clarify this issue, we are interested in the requirements that these theories impose upon their subject (legal practice).

## 2.2 Introducing meaning-use relations

Understanding MURs requires returning to the core idea of the inferentialist project, which is to treat semantics and pragmatics as mutually illuminating, rather than being in conflict or contradictory. This leads to the assumption that the only plausible explanation of how a certain meaning is associated with a vocabulary is to be found in the *use* of this vocabulary (Brandom 2008, 7).[4] In Brandom's words, "[c]laims about the relations between meaning and use have a clear sense only in the context of a specification of the vocabulary in which that use is described or ascribed" (Brandom 1994, xiii). On the grounds of the so-called "semantic pragmatism", these uses are understood as both *practices* that confer the meaning of the vocabulary and as *abilities* whose exercise is constitutive for the fact of "deploying a vocabulary with that meaning" (Brandom 2008, 7). The central interest of this approach is therefore actions, which *count* as expressing the meaning of a certain vocabulary. This can be restated as an attempt to answer the question of what one must do in order to be treated by other participants of the linguistic practice as saying something; as saying that which the applied vocabulary enables one to express. It is also worth observing that this theory, insofar it stresses how these actions *count*, focuses on reciprocal interactions between the agents participating in the practice (the agents acting and the agents that recognise these actions as meaningful). This game-derived nomenclature is characteristic for Brandom's understanding of the correctness of the meaning ascriptions, which is observed from the perspective of the so-called *deontic scorekeeper*, who follows the interactions between participants and assesses how they deploy vocabulary by exercising their abilities (Brandom 1994, 268).[5]

Brandom claims that the assumptions of the mutual relationships between practices, abilities and vocabularies are to be described in the form of basic meaning-use relations (MUR), which are explanatorily prior to introducing more complicated pragmatically mediated relations (which are composed of two or more MURs). The first MUR, called 'practice-vocabulary sufficiency' (PV-sufficiency), obtains between a certain set of practices or a set of abilities (in Brandom's terminol-

---

[4] To make his case about meaning-use relations, Brandom chooses to talk about vocabularies, rather than about words, or terms. We follow this convention here. It seems that choosing the terminology of vocabularies enables Brandom to clearly separate its syntactic and semantic meaning, while being able to maintain the connotation of a vocabulary as a (possibly meaning-conferring) subset of certain language (like a vocabulary of a specific book).

[5] It is worth noting that the perspective of a deontic scorekeeper does not belong to any selected participant of the practice, but can be adopted by anyone who reflects on the inferential features of the practice.

ogy, 'practices-or-abilities'), and a vocabulary which describes how considering someone as applying a vocabulary by other participants of the linguistic practice relies on her participation in certain practices or exercising her abilities:

> **PV-sufficiency**: holds between practices-and-abilities and a vocabulary iff engaging in these practices or exercising abilities suffice to be counted by others as using this vocabulary (Brandom 2008, 9)
>
> Example: the ability to use negation is sufficient to be treated as a competent user of words such as "no" or "not"

Inferentialism assumes that talk about practices-or-abilities is not redundant or trivial only when these practices-or-abilities are understood in terms of the vocabulary which specifies them. For the purpose of this theory, the practices-or-abilities are also relativised to this vocabulary, and their relationship takes form of the next MUR, as follows:

> **VP-sufficiency**: holds between a vocabulary and practices-and-abilities iff this vocabulary suffices to specify this set of practices-or-abilities (Brandom 2008, 8)
>
> Example: the vocabulary of classical logic is sufficient to specify the practice of assigning two truth values (true or false) to any proposition

VP-sufficiency, as a second type of meaning-use relations, makes explicit what one needs to say to specify a certain practice-or-ability (Brandom 2008, 10). What is important to note here is that vocabularies which are VP-sufficient can specify PV-sufficient practices. Therefore, they enable one to distinguish what one must do to count as participating in certain practices (or exercising certain abilities) and in this sense – as deploying certain vocabulary (Brandom 2008, 10). This means that there is a difference between a vocabulary which is VP-sufficient, and a vocabulary which PV-sufficient practices enable one to express. This difference is not, however, merely a difference of explanatory order or a metalevel – for Brandom, the VP-sufficient vocabulary (V') possesses more 'expressive power' than a vocabulary (V) for the deployment of which it specifies sufficient practices-or-abilities (causing a phenomenon which Brandom calls 'pragmatic expressive bootstrapping'; Brandom 2008, 11). This can be understood in terms of an explanatory potential, as V' becomes a metapragmatic vocabulary for V. The relationship between these vocabularies constitutes a proper pragmatically mediated relation, which is understood as a composition of the two previously described basic MUR's (PV-sufficiency and VP-sufficiency):

> **V'V-sufficiency**: holds between vocabulary V' and vocabulary V iff V' is VP-sufficient to specify the practices-or-abilities P that are PV-sufficient to deploy vocabulary V

Example: the vocabulary of mathematics is sufficient to specify practices-or-abilities that are sufficient to deploy the vocabulary of physics

In a simplified version, V'V-sufficiency is described as V' vocabulary's sufficiency to characterise V, and as such V' is considered as a direct syntactic or semantic metavocabulary for V (Brandom 2008, 39).[6] It enables one to grasp how V' is not reducible to V, while allowing one to say what one must do to use V. In this sense, V' (at least pragmatically speaking) provides insight as to the practice of using V, although it is not reducible to V itself. Inferentialism explains V' as explanatorily more powerful, because it expresses what kind of inferences one needs to be able to make in order to be using V.

Apart from claiming that meaning-use relations obtain between practices and vocabularies or between sets of vocabularies only, Brandom introduces yet another basic MUR that holds between practices. Analogously to the relationship of being a pragmatic metavocabulary, an equivalent relation of meta-practices (or meta-abilities) is possible, where the sets of meta-practices (or meta-abilities) elaborate a set of more primitive practices-or-abilities into more complex ones (Brandom 2008, 33). By incorporating these simpler practices-or-abilities as a part of a larger set, the meta-practices have the function of implementing these practices-or-abilities for the performance of more complex tasks:[7]

**PP-sufficiency:** holds between sets of practices-or-abilities iff the capacity to engage in one sort of practice or to exercise one sort of ability is in principle sufficient for the capacity to engage in other practices, or to exercise other abilities (Brandom 2008, 33)

Example: the capacity to use rules of interpretation is sufficient for the practice of interpreting normative acts

When setting forth his proposal of extending the project of analysis to practices of using vocabularies, Brandom introduces – apart from sufficiency meaning-use relations – their corresponding necessity relations. Here, we focus on the necessary re-

---

6 By way of an example, Brandom refers here to the observation made by Huw Price that "although normative vocabulary is not reducible to naturalistic vocabulary, it might still be possible to say in wholly naturalistic vocabulary what one must do in order to be using normative vocabulary" (Brandom 2008, 12). Therefore naturalistic vocabulary has more expressive potential in terms of functioning as a vocabulary which is V'P sufficient to express which practices are necessary to deploy normative vocabulary (even though the latter is not reducible to the former).

7 Similarly to semantic analysis, where the meanings are understood as complex insofar as they are compounded from more basic meanings, analytical pragmatism assumes the complexity of the practices (Brandom 2008, 33).

lationships of two MURs, which hold between sets of practices (PP-necessity) and sets of practices and vocabularies (PV-necessity):

> **PP-necessity:** holds between sets of practices-or-abilities iff it is not possible to engage in or exercise one set of practices-or-abilities unless one also engages in or exercises another (Brandom 2008, 39)
>
> Example: the ability to use variables is necessary for using Euclidean geometry
>
> **PV-necessity:** holds between practices-and-abilities and a vocabulary iff the capacity to say something of a certain kind (that is: to deploy a particular vocabulary), can require being able to do something of a specifiable kind (Brandom 2008, 39–40)
>
> Example: the ability to use the Latin alphabet is necessary for deploying an English vocabulary

The above MURs (or the 'necessity relations') obviously form a stronger relationship between practices and vocabularies, hence providing more insight into the mutual dependence between different sorts of vocabularies and practices.

## 3 The concept of practice in H. L. A. Hart's *The Concept of Law*

Famously, Herbert Hart began *The Concept of Law* with a passage stating that his enterprise can be qualified "as an essay in descriptive sociology" (Hart 1994, v). As Joseph Raz observed, this is no coincidence – one of Hart's greatest contributions to legal theory was to observe that legal rules, like other rules, rest on social sources (Raz 1979, 70). By referring to sociology at the very beginning of his *opus magnum*, Hart reveals that what stands behind his understanding of law is the idea that law is practised and that these practices are intelligible.

Hart's theory has sparked a lively discussion that continues to this day, and it is not possible to fully acknowledge here all the prominent participants of this debate.[8] Much attention has already been given to the semantic side of Hart's enterprise, i.e. its focus on *concepts* and their *meaning* as something static (cf. Dworkin 1986, Stavropoulos 2001, Rodriguez-Blanco 2003), Hart's account of rule-governance (Shapiro 2005, 2011) and his practice theory of rules (Coleman 2001, 2005b, Raz 1999). However, we think that another promising road for analysing Hart's theory –

---

8 Hart's legacy was thoroughly discussed, among others, in the following collections of essays: Hacker/Raz (1977), Coleman (2005a), Kramer et al. (2008).

and a path not yet too well-trodden – is a closer study of Hart's uses of the concept of practice, instead of contributing to the critical assessment of his theory.

Thus, what is practical about law for Hart? It is worth beginning with the observation that Hart uses the term "practice" for the first time in the context of the difference between a habit and a rule. He claims that "[i]f there is to be this right and this presumption at the moment of succession there must, during the reign of the earlier legislator, have been somewhere in the society a general social practice more complex than any that can be described in terms of habit of obedience: there must have been the acceptance of the rule under which the new legislator is entitled to succeed" (Hart 1994, 55).

According to this approach, two types of social practices can be distinguished: (1) "more complex than any that can be described in terms of habit of obedience", and (2) "any that can be described in terms of habit of obedience". We see that within the vocabularies which describe these social practices (at least) three terms are privileged by Hart. When describing the first type, he uses the terms "a rule" and its "acceptance", while discussing the second type, he uses the term "habit" (or the phrase "habit of obedience"). These terms may be considered as *necessary* (for Hart) to specify the *practices-or-abilities* of, accordingly, (1) following the rule or (2) having the habit of obedience. Within the vocabulary of this first-type of social practice there are some word-variations that refer to the social practice, i.e. the "right to specify what is to be done", "standards of behaviour", or the "authority to legislate". Therefore, this vocabulary, along with the rest of the vocabularies of the first type, constitutes the vocabulary which is necessary to specify practice of rule-following ("VP-necessity").

Moreover, Hart would add that there is another practice-or-ability which is necessary to follow any rule, which distinguishes rule-governance from habitual regularities of behaviour: namely the ability to accept this rule from the internal point of view. The so-called internal point of view is crucial for Hart's account of both practices and rules. For Hart, a social rule exists in a community only if members of this community engage in a certain practice from the internal point of view (Shapiro 2006, 1165). As Shapiro rightly observes, the internal point of view plays several roles for Hart's theory: it specifies attitudes that are adopted towards rules (legal rules included), it constitutes one of the existence conditions for social and legal rules; it ensures the intelligibility of legal practice and discourse, and it provides the grounds for the semantics for legal statements (Shapiro 2006, 1158).

As Hacker correctly noted, Hart's account of the internal point of view draws strongly on the Fregean notion of presupposition, interpreted by Strawson (1950) as "a non-logical relation obtaining between putative statements such that the truth of the presupposition was alleged to be a condition of a presupposing statement having any truth value at all" (Hacker 1977, 6). This observation is legitimised because for Hart, "certain types of normative statements are only appropriately used if the speaker has a distinctive attitude towards the norm thus expressed, and the utterance, although it is no part of its meaning that attitude obtains, *presupposes* the existence of the attitude on behalf of the speaker, and hence can be taken to imply (non-logically) that it obtains" (Hacker 1977, 6).

This clearly shows that for Hart the ability to adopt a normative attitude from an internal point of view is necessary to deploy normative statements from the internal point of view, and hence the former is PP-necessary for the latter.

Hart himself identifies the vocabulary that specifies the practice of the internal point of view. He assumes that what distinguishes rule-governance from a mere habit is that only in the case of rules are deviations from a standard considered a criticisable fault, and only in the case of rules is the rule a good reason for making the criticism (Hart 1994, 55). As Hacker puts it, "[i]f social behaviour is to be understood as normative then it must be grasped as being seen by at least some of its participants as conforming to, or deviating from, general standards of conduct" (Hacker 1977, 9). Therefore criticising a deviation from a rule becomes one of the normative attitudes that the practitioners may express towards the behaviour of others, along with the acknowledgement of actions in accordance with a rule and the demands for conformity. Hart calls these attitudes "critical reflective attitudes to certain patterns of behaviour as a common standard" (Hart 1994, 57) which derive from the acceptance of a rule. Such acceptance is nothing other than a "standing disposition of individuals to take such patterns of conduct both as guides to their own future conduct and as standards of criticism which may legitimate demands and various forms of pressure for conformity" (Hart 1994, 255). The practitioner's normative attitudes are manifested by expressing them in normative language, for which Hart gives the following examples: "'I (You) ought not to have moved the Queen like that' 'I (You) must do that', 'That is right', 'That is wrong'" (Hart 1994, 55). Later he explicitly defines this language as "normative terminology of 'ought', 'must', and 'should', 'right' and 'wrong'"(Hart 1994, 57).

Therefore, the normative language is VP-necessary to specify the practice-or-ability of acting from the internal point of view. Moreover, when someone applies normative language, she requires others to treat her as a practitioner acting from the internal point of view, and in this sense her practice becomes intelligible.

There is yet another relationship to be clarified: the relationship between rule-governance and what was so far described as acting from an internal point of view, and participating in the practice or acceptance of a rule. As Hart claims, if rules observed from the internal point of view "are found to exist in the actual practice of a social group, there is no separate question of their validity to be discussed, though of course their value or desirability is open to question" (Hart

1994, 109). This becomes more explicit when Hart explains his practice theory. For him, there is no difference between rules observed from the internal point of view and conventional social practices, if the fact that a group generally conforms to the rule is part of accepting it (Hart 1994, 256). This is further confirmed by Shapiro, who notes that "for Hart, social rules are social practices" (Shapiro 2011, 80). Acting in accordance with a rule from the internal point of view is therefore both PP-necessary and a PP-sufficient practice-or–ability to engage in social practice, because rules and social practices are identical.

Moving on to describe what is specific about legal practice (in relation to other social practices), Hart points out that what underlies "legislative authority" is the same social practice, "in all essentials, as those which underlie the simple direct rules of conduct" (Hart 1994, 58). It has to be underlined, however, that this comment regards the rule of recognition, which is, for Hart, the source of validity of any legal system, from which all the other rules of the system derive their validity (from the fact that they were enacted or accepted according to the rule of recognition). The sovereign is entitled to create rules only if there exists a social practice underlying it, which treats the directives of such legislative authority as binding (Hart 1994, 58; Shapiro 2006, 1165). The question of the validity of law – that is, of the existence and content of the rule of recognition – is a question of empirical fact. Every assertion made by a lawyer regarding the validity of a legal rule presupposes the existence and validity of the rule of recognition in a system, and can be verified by appealing to this fact, that is, to the actual practice of the courts and officials of that system (Hart 1994, 292–293).

As neatly summarised by Shapiro, the "rule of recognition is generated through the convergent and critical behaviour of official identification of certain rules because the rule of recognition is nothing but this practice among officials" (Shapiro 2011, 80). In the *Postscript* to the *Concept of Law*, Hart clarifies that the underlying social practice is only the requirement for the validity of the rule of recognition, while other legal rules only require enactment or acceptance in accordance with the rule of recognition.[9] As M. N. Smith accurately noted "Hart argues that a rule of recognition is at the foundation of law and that the rule of recognition exists only when it is practiced by the relevant officials"

---

9 Hart slightly modified his view in *Postscript*, saying that "[e]nacted legal rules by contrast, though they are identifiable as valid legal rules by the criteria provided by the rule of recognition, may exist as legal rules from the moment of their enactment before any occasion for their practice has arisen and the practice theory is not applicable to them" (Hart 1994, 256). However, we think that this modification does not affect the thesis about the grounds of legislative authority.

(Smith 2006, 266). Smith also captured the precise problem that this "simple" declaration caused for legal philosophers. As he wrote: "[t]he mystery philosophers of law faced, then, was to explain what it meant for a rule of recognition to be practiced by the relevant officials (a gesture in the direction of the internal point of view is quite clearly insufficient). To resolve this mystery, some philosophers appeal to Lewis's account of conventions or to Bratman's account of shared agency." (Smith 2006, 266).

From this it follows that Hart draws a relation of practice-practice necessity (PP-necessity) between the social practice of the legal officials applying rule of recognition and the practice of the legislator (Hart's "Rex"). This means that the officials' participation in the legal practice (on the grounds of a relevant system) constitutes a necessary meta-practice for the participation in the practice of legislation. The former also allows its practitioners to count as *deploying* the vocabulary of Rex's legislation. Therefore, there is – in Brandomian terms – a relation of practice-vocabulary necessity between the practice of the legal officials and the vocabulary of Rex's legislation (PV-necessity).

The practice theory of law remains the central area of Hart's use of the notion of practice in his legal theory. Despite using the concept to describe some selected types of practices, i.e. when he speaks about a "customary practice" (Hart 1994, 95), the "practice of religion" (Hart 1994, 183) or "a traditional practice" (to address Guy Fawkes Night celebrations), he remains concerned with the social practices that ground the validity of law. However, a peripheral yet overlooked aspect of his use of the concept is that Hart observes that mutual relations exist between practices, such as when a legislative practice can decay a moral standard or repeal tradition.

## 4 The place of the practices in "Law's Empire"

Ronald Dworkin famously claimed that:

> [g]eneral theories of law, like general theories of courtesy and justice, must be abstract because they aim to interpret the main point and structure of legal practice, not some particular part or department of it. But for all their abstraction, they are constructive interpretations: they try to show legal practice as a whole in its best light to achieve equilibrium between legal practice as they find it and the best justification of that practice. (. . .) Law cannot flourish as an interpretive enterprise in any community unless there is enough initial agreement about what practices are legal practices so that lawyers argue about the best interpretation of roughly the same data. (Dworkin 1986, 90).

The following commentary focuses mainly on how Dworkin, in his criticism of Hart's approach, came to the conclusion that an abstract theory of law is necessary in order to interpret legal practice. In other words, what we are concerned with here is precisely uncovering the curious relationship that Dworkin outlined between the theory of law and law as a practice. As a consequence, he postulated a specific way of researching law.

As Dworkin commented in *Taking Rights Seriously*, one of the possible sources of a rule's authority for H.L.A. Hart is "(. . .) that [a certain] group through its practices *accepts* the rule as a standard for its conduct. (. . .) A practice constitutes the acceptance of a rule only when those who follow the practice regard the rule as binding, and recognize the rule as a reason or justification for their own behavior and as a reason for criticizing the behavior of others who do not obey it" (Dworkin 1978, 20). The second source of a rule's authority is its validity. However, as Dworkin argues, it is impossible to preserve the Hartian distinction between sources of a rule's authority, i.e. (1) the practice of acceptance and (2) the validity of rules. This criticism results from the facts that (i) there are practices supported by principles (which are logically distinct from rules, because principles are not applicable in an all-or-nothing fashion), (ii) and that those principles, in contrast to the rules, are open to being challenged. The example of principles unveils that the link between principles and the practice of officials, whose actions amount to the master rule of recognition (in Hart's terms, granting the validity of the system), is insufficient (Dworkin 1978, 41). As a consequence, Dworkin refuses to treat principles as something ultimate. In this sense, he rejects "the positivists' first tenet, that the law of a community is distinguished from other social standards by some test in the form of a master rule", because as he argues, this is due to the nature of the principles: their controversiality, numberlessness, different weights, and changeability. (Dworkin 1978, 43–44).

Dworkin's criticism was directed against the concept of law as a system of rules. In contrast, the concept of the principle was crucial for his argument. At first glance, it may seem that Dworkin's criticism of the social roots of rules – along with introducing the notion of the principle – would entail abandoning (at least to some extent) of the concept of practice in Dworkin's theory. As it turns out, he in no way rejected the importance of the concept of practice for building his own theory of law. In Dworkin's legal theory, this concept is used to legitimise the "upper layer" of this theory, that is, the interpretative practice in which we engage our value-based reasonings.

Dworkin describes law as "an interpretive concept". The interpretation is provided by judges, who "(. . .) normally recognize a duty to continue rather than discard the practice they have joined. So they develop, in response to their own convictions and instincts, working theories about their own convictions and instincts, working theories about the best interpretation of their responsibilities under that practice" (Dworkin 1986, 87). The practice-to-the-best-interpretation-of-legal-practice (for short: "first-order practice") seems to be somehow distinct from the legal practice itself ("second-order practice"). As with the practice of courtesy in Dworkin's famous example, it is assumed that the first-order practice has value (or as Dworkin also put it, "serves some interest or purpose or enforces some principle", "has some point") and the rules must correspond to this point (Dworkin 1986, 47).

At least one obvious objection arises against this *prima facie* distinction: is it really Dworkin's position that these are two different practices? Or, to put it in a different way, maybe the first-order practice is identical to the second-order one? We think that one can establish two arguments for this differentiation, which are – in fact – two sides of the same coin. The first argument is based on the temporal sequence of the practices. The second-order practice is always a matter of the past. Since it already belongs to the past, we can only provide a certain flashback from this practice. Inversely, the first-order practice in this view is always something actual, and – according to Dworkin's view on the value of the practice – is forward-looking. The second argument – which can be put forward on the opposite side of analysed "coin" – refers to the attitudes that the participant may adopt towards the practice. In the case of the second-order practices, there is no possibility of "internally" participating in the practice. The flashback is made outside of the practice, as in the case of the interpretation made in the course of the first-order practice.

A second objection may now arise: is a first-order practice really a practice or, in other words, are there any norms that govern it? And the answer, following Dworkin's approach, must be "yes". If the first-order practice has value, as Dworkin holds, this assumption can be reformulated into the following rule:

> **(R1)** The interpreter X is obligated to provide the interpretation of legal practice $I_{LP}$, which is compatible with the value of this practice $V_{LP}$.

> This rule has to be supplemented by Dworkin's assumption of "the best interpretation". After doing so, the rule can be rephrased as the following one:

> **(R2)** The interpreter X is obligated to provide the best interpretation of legal practice $I_{LP}$, according to the value of this practice $V_{LP}$.

Finally, is the first-order practice a legal practice? Let us briefly sketch what the stages of first-order practice are for Dworkin. First of all, that practice has a "pre-interpretive" stage, where the participants identify specific rules and standards of the second-order practice. The second stage is called "an interpretative" stage, and this is where the interpreter sets forth the justification of the main components of the second-order practice. And finally, Dworkin distinguishes third stage of the first-order practice, named "postinterpretive" or "reforming". At this stage, the interpreter "adjusts his sense of what the practice »really« requires so as to better serve the justification he accepts at the interpretive stage" (Dworkin 1986, 66).

It might be said that the sufficient criterion for qualifying the first-order practice as legal is that the participants of this practice are "public officials but whose decisions also affect the legal rights of their fellow citizens" (Dworkin 1986, 12). Dworkin's approach is in this sense pragmatic (or it seems to be pragmatic) in that he takes into account the different attitudes of judges. As he himself notes: "My project is narrow in a different way as well. It centers on formal adjudication, on judges in black robes, but these are not the only or even the most important actors in the legal drama. A more complete study of legal practice would attend to legislators, policemen, district attorneys, welfare officers, school board chairmen, a great variety of other officials, and to people like bankers and managers and union officers (. . .)" (Dworkin 1986, 12). His criticism of various philosophical positions is also personified. We see that Dworkin labels judges (e.g. "a conventionalist judge", Dworkin 1986, 133; "an imaginative pragmatist judge", Dworkin 1986, 155; or "Hercules" and "Hercules' practice", Dworkin 1986, 250). Therefore, the first-order practice can be qualified legitimately as legal due to the status attributable to its practitioners.[10]

Therefore, the conclusion stemming from Dworkin's *Law's Empire* seems confusing, for it suggests that the practice of the best interpretation should be considered necessary for engaging in legal practice, or that – in other words – it is not possible to engage in legal practice unless one also engages in the practice of the best interpretation. However, if the practice of the best interpretation turns out to be legal practice, it cannot therefore be seen as necessary for itself.

It is only at this point that the clear distinction between first-order and second-order legal practices collapses. Legal practice for Dworkin is *always* the practice of the practice of flashbacks – as we described it above. This practice involves establishing interpretations of the past decisions of relevant legal officials. Therefore, it requires a more basic ability to interpret the past practice, which is a necessary ability to *make* an interpretation (PP-necessity). As it follows from Dworkin's claims, the practice has a value, so providing a proper interpretation also requires the ability to recognise the value of the practice (PP-necessity). On the other hand, to specify legal practice we have to engage the vocabulary of principles, values, and interpretations (VP-necessity). The meta-practice-or-ability is, in fact, the practice-or-ability of recognising relevant values, and the practice of the lower level is the legal practice itself,

---

10 On the institutional status of agents see also Dybowski (2018, 45).

understood as the practice-of-best-interpretation-of-(past)-legal-practice. To conclude, the meta-practice is necessary for the lower level one, and the meta-practice does not have to be qualified as legal. This is because, in the Dworkinian view, legal practice is only one of the many social practices in which we engage the more fundamental, non-legal practice-or-ability of value determination.

# 5 The practice of law: J. Raz on the nature of practices underlying law and value

Raz seems to be the most explicit about his understanding of the concept of practice when he speaks about values. However, he mentions practices in several other places: that is, when speaking about the existence and validity of law, about legal rules, and about legal interpretation. This clearly shows how important this concept is for his general outlook on law itself. Moreover, combining his proposal to define values through their underlying practices with his standpoint on law's own virtue leads to a more thorough understanding of his views on the rule of law, too. The discussion of practices permeates Raz's thinking about law, which seems symptomatic for the post–Hartian positivists' theories of law. However, Raz's unique theory of value (apart from many other reasons) justifies a separate analysis through the adopted lens.

Because the clearest view on the concept of practice seems to emerge from Raz's lectures on moral theory,[11] we will start from analysing this usage – thus from the perspective of ethics, rather than law. As was already implied, this can be taken to be of significance for his understanding of law as well. This caveat seems necessary, given Raz's firm position on the relationship between law and morals (and the lack thereof), which is defining for his exclusive positivist standpoint.

For Raz, the social dependence of value hinges on two theses. First, that values exist only if there are (or were) social practices sustaining them. Second, that all values (with some exceptions) rely on social practices directly, or because they depend on the existence of other values that are directly dependent on social practices (Raz 2003, 116). The vocabulary Raz uses to describe the necessary conditions for the emergence of a practice is that of beliefs: a population sustains the practice underlying a value iff people act as they would, were they aware of the

---

[11] The Tanner Lectures delivered by Joseph Raz at Berkeley in 2001, later published by OUP in 2003 as the "Practice of value". The numbering and date indicated in commas refer to the book.

practice and shared the belief that it is worthwhile to conduct themselves the way they do so.

This picture allows that the practitioners, even though they engage in the practice, may remain unaware of the exact value they are sustaining, or that they mistake the value for something else (although it leads them to the same conduct that acting knowledgeably in accordance with the value would lead to). However, conduct merely imitating actions sustaining value cannot be treated as coincidentally practising value. While the practices "entail common knowledge of their terms, i.e., of what they require, we need not expect the practices to be informed by a good understanding of the values that could justify or make sense of them" (Raz 2003, 117). Yet values are not only practised by actions or conduct – more general values are put into practice and sustained by specific ones. Here, Raz gives the example of the respect for freedom, which is sustained by adhering to the rule of law (Raz 2003, 117–118). In the above sense, the doxastic vocabulary (that is, the vocabulary of forming beliefs) is VP-necessary for expressing what counts as practising specific values. On the other hand, the practice of specific values can be PP-necessary for practising more general values. Moreover, the doxastic vocabulary of forming beliefs is V'V-necessary for the vocabulary of the more general values, because it specifies practices VP-necessary to specify these practices – that is, the practising of specific values – those which are PV-necessary to determine the content of the general values.

It is interesting to note that the ontological relationship between different types of value seems different to that of vocabularies. Raz claims that a sustaining practice is only necessary for the emergence of a value, not for its subsequent upholding. After the practice is established, it becomes only a non-necessary condition for an existing value, as the practices do not have to persist as long as the value does (Raz 2003, 118).

Although Raz claims that both the goods and the bads require social practices to sustain them, it is neither the case that these practices are assessed as good or as bad based on the beliefs of the practitioners, nor that just any kind of practice leads to the emergence of value. This is because values can only be described in normative, evaluative terms, and their further assessment as good or bad cannot fully account for the character of value. Moreover, for Raz concepts of false value cannot have instances – "if there is no value X then the concept of X is a concept of a false or illusory value and there is nothing that can have the value X", and thus they fail to explain anything in the world, despite being used to do so (Raz

2003, 120).[12] These are the arguments underlying Raz's non-reductionist[13] account of value. What they imply is that the relationship between practices and the normative beliefs of the practitioners is content-independent. The practitioner's beliefs about the worthwhile character of their actions lead to sustaining practices regardless of their truth-value, that is, despite being wrong about this quality. Moreover, social practice can amount to a variety of goods or values (Raz 1999, 195). Value depends, therefore, on the concept of value that the practitioners have, and on the fact that they can understand something about the nature of a value (Raz 2003, 128). On the other hand, a value depends on the content of the practice, as the value requires valuers for its emergence (Raz 2003, 123). This is not to say that just any kind of valuer will suffice. What is constitutive of value is a valuer who is not mistaken about the value, which means that she can engage with the value and appreciate it (Raz 2003, 125).

Raz's reflections about values are important for the value of law, too. Explaining the genre-based approach to defining the concept of value – which enables identifying valuable objects and assessing a value by the standards of that to which the value belongs – Raz gives an example of the excellence of law qua law which derives from the rule of law or from actually possessing the authority law claims to have (Raz 2003, 126). In other places, he explicitly defines the rule of law as law's own virtue, the value law should possess (Raz 2018, 1). When Raz speaks about the nature of the state it seems that the relationship between a value and a valuable object is even stronger, as the standards of excellence of a state (a creature of law) belongs to its concept (as a value is an object of the state's claim). This may be connected with the claim that genre defines the structure, or combination, of the standards of excellence for a value, which derive from its nature. Therefore, it can be said that assessing the value or disvalue of law requires certain skills – the ability to engage with the value and knowing the structure of standards that in virtue of law qua law amount to its excellence. This, in turn, enables one to say something about the legal valuers. Adopting normative attitudes towards legal actions involves knowledge as to the requirements of the rule of law and, consequently, as to the concept of law. It should be noticed that this knowledge does not concern the content of law, but the procedural aspects of its generation and application (Raz 2018, 2). This conclusion is validated by the fact

---

[12] This shows how Raz's ontological thesis is intertwined with the epistemological one – values can exist only if it is possible to understand something about their nature, and therefore – to have access to them. This also connected to the structure of value, which according to Raz is genre-based and derives from a combination of qualities, or practices.

[13] Non–reductionist in the sense that the content of value is not reducible to the contents of the practitioner's beliefs.

that in some places Raz uses the concept of a shared understanding when speaking about social practices. He does this to underline that they should be distinguished from reasoned regularities (Raz 1999, 199).

Knowledge of the specificity of a legal genre – thus regarding its standards and structure – becomes a PP-necessary practice-or-ability for the emergence or sustaining of law's value. In other words, the ability to use the concept of law is a practice-or-ability PP-necessary for adopting normative beliefs, and hence for the action that leads to sustaining the value of a legal practice. This can be understood as connected with the Hartian assumption that the social practice constitutive for legal practice is the practice undertaken by legal officials on whose normative attitudes depends whether the law claims the authority to guide the behaviour of its subjects. Raz agrees with Hart that for the law to remain efficacious "conformity is not enough", and that the acceptance of law at least on the part of legal officials is necessary (Raz 1997, 210).

Yet Raz's standpoint on the existence of values cannot be treated as directly transferable, by analogy, to his viewpoint on the existence of law. The analogy works if the law can be considered as value generating – if it actually has the rule of law, not if it only claims it. From the Razian account of value it does not follow, of course, that every legal practice is value-sustaining or that the existence of law depends on its value, nor that abiding by the rule of law guarantees that the law is good or just (Raz 2018, 10). The rule of law is only a necessary condition of law's virtue, not a sufficient one. Nevertheless, Raz admits that practices are a part of what the law consists of, as he claims that "[t]he Law is a structure of rules, institutions, practices and the common understandings that unite them, which normally are an aspect of some social organisation: state, city, university, corporation" (Raz 2018, 1).

In *Practical Reason and Norms* Raz declared that "the law is a normative system claiming authority that is both comprehensive and supreme" (Raz 1975, 149). This system, as he describes it in *The Concept of a Legal System*, can be conceived of as a system of reasons for action. The legality of these reasons depends on two main conditions: that these reasons are applied and recognized by the system of courts, and that those courts are bound to apply them, according to their practice and customs (Raz 1997, 212). The legal reasons are not addressed to the courts only, but to all kinds of agents. However, it remains their necessary condition that "the courts are bound to recognize them and to draw appropriate conclusions from conformity or non-conformity with them" (Raz 1997, 215). This draws attention to the relationship between reasons and rules, and between rules and practices in Raz's account. These relationships can be described in relation to the role of reasons and rules in practical reasoning. Raz says that rules are mediators between deeper-level considerations and concrete decisions, hence they are an intermediate level of reasons to which one appeals where a need for a decision arises (Raz 1986, 58). They provide grounds for fruitful reasoning (finalised with a decision) because they enable committing to the sets of actions instead of performing a case-by-case analysis. If they are shared by a community, the rules become common standards of conduct, i.e. shared practices, which are necessary for the existence of an orderly society (Raz 1986, 58). This leads to the conclusion that the vocabulary of reasons is VP-necessary for specifying the ability to participate in a social practice and being guided by a rule. The vocabulary of the practical reasoning (reasons) is also V'V– necessary for the vocabulary of social practices, because it enables one to express what one needs to do in order to deploy the vocabulary of these practices. In the case of legal rules, their source is the practice of the courts, but their relationship to this practice differs in the case of the ultimate and ordinary legal rules (which together constitute the set of all legal rules). The shared practice of the courts is a constitutive proof that a rule is a legal rule, but it is not the reason for its validity nor the reason to apply it. Conversely, the ultimate legal rules are the grounds of validity of ordinary legal rules, therefore all the legal rules are either grounded in social facts (practices) or in other legal rules, which are themselves grounded in social facts (Raz 1979, 68–69).

Apart from addressing the practice of law broadly, that is, as a general term to describe different specific practices that law incorporates, Raz sometimes seems to refer to practice in individual terms, i.e. as a specific practice of adjudication (Raz 2011: 298), or when he speaks about legal interpretation as a certain practice of law that can be accessed via legal documents (Raz 2018, 11). He speaks about the practice of entrenching rights in similar terms (Raz 1986, 263).

It may be concluded that with regard to both values and law, Raz's account is unified on the point of their dependence on social forms that take the form of sustaining practices. In the case of law, of key importance is the social status of the practitioners – that is, legal officials – while in the case of values what is crucial are the beliefs about a certain quality of a behaviour. Raz admits that the existence of law is ultimately based on social practices but denies that this fact detracts from their normativity (Raz 1979, 89).

# 6 How to make theory of law practical?

The last theory covered by our analysis is George Pavlakos' Practice Theory of Law, which is included mainly because of its unusual conception of practice. This author proposes to understand law as a *constraint-generating concept* (Pavlakos 2007, 211). These legal limitations arise as agents make moves in legal practice.

Unlike the other authors mentioned here, Pavlakos' effort is focused on providing such characteristics of legal practice that it can be considered as a "platform" for the creation of legal knowledge. Therefore, following pragmatic rationalism, the author is able to explain the generation of knowledge in practical terms: as connected "with the idea of a practice of judging which is normatively constrained by reasons or, to say the same thing in different terms, reflexive" (Pavlakos 2007, 239). Thereby, he tries to expose that law is necessarily connected with every other practical domain.

Pavlakos discusses different approaches to the so-called *mind-world dualism*. This kind of dualism forces those who want to avoid it to seek intermediaries between the mind and the world. Therefore, if practices are to be regarded as existing outside dualism, they cannot be such intermediaries. He follows a specific version of a pragmatic reading of Wittgenstein's works, where meaning and understanding are placed within the context of different agents' actions. Let us look at the main theses of this pragmatic account of practices.

First, it is claimed that this approach is in line with the argument that *semantics exhausts ontology* (Pavlakos 2007, 39), and (a bit further) that *"grammar has an explanatory priority over reference and not that reference is non–existent"* (Pavlakos 2007, 54–55). Grammar is a concept borrowed from Wittgenstein's philosophy. It has a regulative function with respect to our thinking, and, what is crucially important here, it regulates thinking in the normative way. Therefore, the practice Pavlakos has in mind is definitely linguistic.

Second, he tries to secure the objectivity of practices and calls them *ratification-independent patterns* (Pavlakos 2007, 112). The Practice Theory of Law rejects practices as self-interpreting entities. Pavlakos claims that these practices "(. . .) lie on the same level as either minds and world and, therefore, are common to all minds (. . .) practices do not need a second- (or third-, fourth- . . . or vn-) order interpretation (or any interpretation at all, which is the same as saying that they come already interpreted) in order to constitute a common ground for all minds. It follows that practices do not succumb to an infinite regress but are ratification-independent, ie objective" (Pavlakos 2007, 112).

However, how are ratification–independent practices possible? Pavlakos proposes the following conditions. A practice must be held by something that is within the practices. Such constraints are equated with *reasons* for obeying the rules of practice, and the reasons, in turn, are regarded as facts. Those reasons, however, must meet three further conditions: (1) the condition of *independent identifiability*, (2) the condition of *direct readability*, and (3) the condition of *fallibility of reasons* (Pavlakos 2007, 133–134).

As Pavlakos claims, this approach reveals that practice is not a mechanical enterprise; it has a reflective dimension. The participants of the practices are endowed with the opportunity to critically evaluate individual cases of application of the rule. In this way, the practice also allows for understanding who its participants are, revealing its personal character (Pavlakos 2007, 135). He also distinguished "between two levels of judging: a fundamental one which relates to grammar and another one that relates to all other practices of judging: (. . .) *judging simpliciter* (JS) and *judging in a domain* (JD) respectively" (Pavlakos 2007, 137). Such a description of practices leads the author to conclusions about the *"depth of practice"*. This depth is the subject's ability to present the reasons underlying the practice, while grammar in this context allows for the "objectification" of reasons as the basis of our justifications. Once the reasons are "objectified" inside cognitive practice, the possibility of invoking them in order to justify judgments opens up (Pavlakos 2007, 139).

Following Kant, Pavlakos reconstructs the concept of a person, and concludes that (1) it is not possible to draw strict boundaries between the different areas of judgement,[14] and that (2) the concept of the person guarantees the so-called Unity of Practical Reason[15] (Pavlakos 2007, 149–151). Therefore, he accepts that there is a shared level (called the "practice of universalisation") for all practical reasons. This is the level where we see the connections between these reasons. We can generate – as Pavlakos claims – the reasons through different social practices. Nevertheless, this procedure does not deprive what is generated of its "common denominator".

Pavlakos criticises both Hart's legal conventionalism and Dworkin's interpretivism. As he notes:

> (. . .) there is an important difference between conventionalism and interpretivism: in the former, legal norms are being reduced to behavioural facts about those who partake of the practice of law. In the latter, legal norms refer to moral commands that do not depend on the practice of law but, conversely, are called upon to 'shape' or 'guide' the cognition this practice effects. (. . .). Such differences notwithstanding, both failures have one in common: they fail to account for our knowledge of the law (Pavlakos 2007, 208).

---

**14** In the weaker version, each subject making an act of judging is sooner or later obliged to conduct moral argumentation, or, in the stronger version, the autonomy of persons is presented as dominant in the sense that the actors retain their moral status as a person in various areas of judgement (see Pavlakos 2007, 149–150).

**15** As Pavlakos claims "[t]he explication of normative reasons, as resting on public practices of universalisation, guarantees the continuity of practical reason. Although practical reasons are generated in different contexts by distinct practices (law, morality, ethics, etc) the idea of universalisation helps to bring out the rationale which is common to all of them. Thus, despite differences between various types of practical reasons (be they legal, moral or otherwise practical) it is possible to relate them to one another as a result of their being instances or outcomes of a practice of universalisation" (Pavlakos 2007, 151).

Pavlakos links the possibility of having knowledge with the ability to denote objects. In legal practice such objects are legal norms (as non-linguistic entities), expressed by linguistic entities. As he claims, legal norms are "higher-order abstract entities or facts which are available to lawyers' cognitive powers" (Pavlakos 2007, 230). Understood in this way, legal norms are seen as reasons that constrain legal practice (see Pavlakos 2007, 224 and 231).

This short recapitulation of the main tenets of George Pavlakos' theory allows us to engage the metatheoretical tool we proposed at the beginning of the paper. As we can notice, legal knowledge requires the practice of law to arise. However, every *judgement in a domain* (such as law), is connected with a fundamental one, which is *judgement simpliciter*. As Pavlakos noted "the discursive character of law disclosed a fundamental practical commitment of legal practice: in being a practice of judging, legal practice must respect a certain amount of autonomy of agents, which is enshrined in the activity of reflexive judging" (Pavlakos 2007, 240). Therefore, it is hard to say that this practice is sufficient for the generation of this kind of knowledge on the grounds of the Practice Theory of Law. Legal knowledge is composed of claims about legal norms (Pavlakos 2007, 239). And having knowledge about legal norms seems to be intertwined with being able to participate in other practices such as morality or ethics (due to their "common denominator", which is the practice of universalisation).

Pavlakos suggests that the correctness of using legal reasons (which can be represented as linguistic entities) is determined within the legal practice. Therefore, the relation of PV-necessity exists between the vocabulary of law (the vocabulary of linguistic entities which represents legal norms) and legal practice. Moreover, it is also postulated that there is at least one relation between the general practice of judging autonomously (P') and legal (as well as other similar) practice (P). We think that this is a relation of P'P-necessity. It was already implied that participation in legal practice (P) is also connected with such abilities as the ability to present the reasons underlying the practice ($P_1$). Therefore, another relation $PP_1$-necessity can be distinguished here. On the other hand, Pavlakos tries to show that the vocabulary of reasons, and norms, as well as the domains which he recognizes as connected with them, are sufficient to specify legal practice (VP-sufficiency). In this context, the question about legal knowledge becomes only a pretext to say that although it is "technically" possible to distinguish a legal practice and its components, this practice is "indeed" inextricably linked with the more fundamental activities of the entities who participate in it.

# 7 Conclusions

First of all, what follows from the above analysis is that the concept of practice is deployed by different legal theorists to solve different problems regarding law. For Hart, this is the solution for the question of the existence of rules and rule-governed behaviour. Dworkin applies the concept of practice to account for the legitimacy of legal decisions, and to establish the role of non-legal reasoning in achieving the best interpretation of legal practice. For Raz, practices explain the grounding of values and legal rules, and are a main tenet in describing the role of the latter in practical reasoning. Finally, we observed that George Pavlakos changed the explanatory direction of using the concept of practice. These authors used the concept of practice to explain the practical aspect of the selected concepts (Hart's Practice Theory of Rules, or Raz's Practice of Value). On the other hand, Dworkin identified the solution to the problem of the legitimacy of legal decisions in a selected practice, i.e. the practice of interpretation. It seems that Pavlakos' theory does not treat the concept of practice as an explanatory tool. Instead, as we have indicated, the concept of legal knowledge becomes for him only an impulse to characterise the essence of legal practices.

Second, we aimed at establishing the methodological consequences of commitments to the selected vocabularies. We have already discussed two of the three questions we asked at the beginning of this paper, i.e. (1) what one must *do* (in terms of *practices-or-abilities*) in order to count as *being* part of the practice in scope, and what these *practices-or-abilities* allow the practitioners to express, and (2) what the *vocabulary* is which *specifies* those *practices-or-abilities*. Now, we have to address the last one, concerning the methodological consequences of committing oneself to the specific vocabularies while describing legal practice.

The selected legal philosophers commit themselves (at least implicitly) to using different relations of necessity when building their legal theories. It can be observed that the apparatus we use reveals a certain common feature of the description of law as a practice. In each of the analysed approaches, three of the four of the theorists do not commit to the sufficiency relationship between vocabularies or practices. The exception is the VP-sufficiency relation, which is possible to distinguish when analysing Pavlakos' Practice Theory of Law. This relationship occurs between the vocabulary of reasons, and norms (as well as the domains which Pavlakos recognised as connected with them), which are sufficient to specify legal practice. This conclusion derives from the theoretical objective Pavlakos seems to adopt, namely to establish the minimum criteria for identifying legal practice.

We think that the reason for committing to necessity relationships may derive from the fact that legal practice does not constitute an autonomous practice. The indicated philosophers postulate that legal practice must draw from other, more primitive practices. Hence this limits the application of the sufficiency relationship to describe legal practice. Consequently, legal philosophers seem compelled to reflect on certain legal necessities. Therefore, the differences between the selected theories of law as to the necessity-relationship concern mainly the following: what vocabularies are necessary to characterise legal practice, or what other practices-or-abilities are necessary for agents to be counted as legal practitioners.

Third, we promised to provide some general arguments as to why the applied MUR relationships may become an attractive methodological tool for further analyses of legal theories. We believe that the main advantage of the inferentialist apparatus is that it focuses on the relations between concepts. This leads to more complex analyses of pairs or sets of concepts rather than selected concepts only. Moreover, it enables analysis to be oriented towards the agents whom the theories concern. The proposed relational analysis requires observing the social aspects of applying concepts, by focusing attention on the abilities and actions of the subjects of social theories. Another advantage could be that the inferentialist's analysis enables one to assess the degree to which the selected concept (in our case, the concept of practice) draws on other key concepts for a selected type of theory. In the case of legal theories, these are, among other things, rules, obligations, interpretation, norms, legislature, statutes, officials and institutions. Such optics unfold whether the analysed correlations between concepts are central for a theory, rather than peripheral.

# Bibliography

Brandom, Robert B. 1994. *Making It Explicit. Reasoning, Representing, and Discursive Commitment*, Cambridge (Massachusetts), London, Harvard University Press.
Brandom, Robert B. 2000. *Articulating Reasons. An Introduction to Inferentialism*, Cambridge (Massachusetts), London, Harvard University Press.
Brandom, Robert B. 2008. *Between Saying and Doing. Towards An Analytic Pragmatism*, Oxford, Oxford University Press.
Coleman, Jules. 2001. *The Practice of Principle. In Defence of a Pragmatist Approach to Legal Theory*, Oxford, Oxford University Press.
Coleman, Jules (ed.). 2005a. *Hart's Postscript. Essays on the Postscript to the Concept of Law*, Oxford, Oxford University Press, 99–148.
Coleman, Jules. 2005b. *Incorporationism, Conventionality, and the Practical Difference Thesis*. In Jules Coleman (ed.), *Hart's Postscript. Essays on the Postscript to the Concept of Law*, Oxford, Oxford University Press, 99–148.
Dworkin, Ronald. 1978. *Taking Rights Seriously*, Cambridge (Massachusetts), Harvard University Press.
Dworkin, Ronald. 1986. *Law's Empire*, Cambridge (Massachusetts), London, Harvard University Press.

Dybowski, Maciej. 2018. *Articulating Ratio Legis and Practical Reasoning*. In Verena Klappstein & Maciej Dybowski (ed.), *Ratio Legis. Philosophical and Theoretical Perspectives*, Cham, Springer, 29–55.
Hacker, Peter M. S. 1977. *Hart's Philosophy of Law*. In Peter M.S. Hacker & Joseph Raz (ed.), *Law, Morality and Society. Essays in Honour of H. L. A. Hart*, Oxford, Oxford University Press, 1–25.
Hacker, Peter M.S. & Joseph Raz (eds.). 1977. *Law, Morality and Society. Essays in Honour of H. L. A. Hart*, Oxford, Oxford University Press.
Hart, Herbert L. A. 1994. *The Concept of Law*, Oxford, Oxford University Press.
Kramer, Matthew, Claire Grant, Ben Colbrun & Antony Hatzistavrou (eds.). 2008. *The Legacy of H.L.A. Hart: Legal, Political and Moral Philosophy*, Oxford, Oxford University Press.
Pavlakos, George. 2007. *Our Knowledge of the Law. Objectivity and Practice in Legal Theory*, Oxford and Portland (Oregon), Hart Publishing.
Pelc, Jerzy. 1982. *Wstęp do semiotyki*, Warszawa, Państwowe Wydawnictwo "Wiedza Powszechna".
Price, Huw. 2004. *Naturalism without Representationalism*. In Mario de Caro & David Macarthur (ed.), *Naturalism in Question*, Cambridge (Massachusetts), London, Harvard University Press, 71–88.
Raz, Joseph. 1975. *Practical Reason and Norms*, London, Hutchinson.
Raz, Joseph. 1979. *The authority of law: Essays on law and morality*, Oxford, Oxford University Press.
Raz, Joseph. 1986. *Morality of Freedom*, Oxford, Oxford University Press.
Raz, Joseph. 1997. *The Concept of a Legal System. An Introduction to the Theory of Legal System*, Oxford, Oxford University Press.
Raz, Joseph. 1999. *Engaging Reason. On the Theory of Value and Action*, Oxford, Oxford University Press.
Raz, Joseph. 2003. *The Practice of Value*, Oxford, Oxford University Press.
Raz, Joseph. 2011. *Between Authority and Interpretation. On the Theory of Law and Practical Reason*, Oxford, Oxford University Press.
Raz, Joseph. 2018. *The Law's Own Virtue*, Columbia Public Law Research Paper No. 14–609, Oxford Legal Studies Research Paper No. 10/2019, King's College London Law School Research Paper No. 2019-17, Available at SSRN: https://ssrn.com/abstract=3262030 or http://dx.doi.org/10.2139/ssrn.3262030.
Rodriguez-Blanco, Veronica. 2003. *A Defence of Hart's Semantics as Nonambitious Conceptual Analysis*, Legal Theory 9, 99–124.
Shapiro, Scott. 2005. *On Hart's Way Out*. In Jules Coleman (ed.), *Hart's Postscript. Essays on the Postscript to the Concept of Law*, Oxford, Oxford University Press, 149–192.
Shapiro, Scott. 2006. *What Is the Internal Point of View?*, Fordham Law Review 75, 1157–1170.
Shapiro, Scott. 2011. *Legality*, Cambridge (Massachusetts), London, Harvard University Press.
Smith, Matthew Noah. 2006. *The Law as a Social Practice: Are Shared Activities at the Foundations of Law?*, Legal Theory 12, 265–292.
Stavropoulos, Nicolaos. (2001) *Hart's semantics*. In Jules L. Coleman (ed.), *Hart's Postscript: Essays on the Postscript to 'the Concept of Law'*. Oxford University Press, p. 59–98.
Stavropoulos, Nicolaos. 2005. *Hart's Semantics*. In Jules Coleman (ed.), *Hart's Postscript. Essays on the Postscript to the Concept of Law*, Oxford, Oxford University Press, 59–98.
Strawson, Peter Frederick. 1950. *On referring*, Mind 59, 320–344.
Tamanaha, Brian Z. 1999. *Realistic Socio-Legal Theory: Pragmatism and A Social Theory of Law*, Oxford, Oxford University Press.
Wittgenstein, Ludwig. 1953. *Philosophical Investigations*, Oxford, Basil Blackwell.

Mateusz Zeifert
# Natural semantic (legal?) metalanguage. What can legal theory learn from Anna Wierzbicka?

**Abstract:** Natural Semantic Metalanguage (NSM) is a semantic theory originated by Anna Wierzbicka. It provides a list of "semantic primes" – concepts that are claimed to be primary (i.e. they cannot be explained in simpler terms) and universal (i.e. are lexicalised in all human languages). They offer a unique tool for a cross-linguistic and cross-cultural analysis of meaning. The paper's thesis is that NSM may prove useful in legal contexts. Several possible areas of application are identified. Firstly, NSM could enhance the comprehensibility of legal texts, which are notoriously difficult for laypeople to read. Secondly, NSM could be used for semantic analyses of legal terms, which typically lack coherent methodology. Thirdly, NSM could provide a much-needed common point of reference (or *tertium comparationis*) in comparative law. Fourthly, NSM could help draft multilingual and culture-neutral documents in international law.

**Keywords:** Natural Semantic Metalanguage, legal language, legal semantics, plain legal language, comparative law

## 1 Introduction

There should be nothing controversial in saying that law is a linguistic device and that theories about language constitute an important source of inspiration for legal scholars. The most famous example is probably Herbert Hart's theory of open texture, which was inspired by the ideas of several prominent philosophers: Friedrich Weissmann, Ludwig Wittgenstein and John L. Austin (Hart 2012; Zeifert 2022, 412–414). Although its scope and significance are still the subject of heated academic discussion (Endicott 2008; Stavropoulos 2001; Müller Fonseca 2018), it has undoubtedly contributed to our understanding of legal language, statutory interpretation and the concept of law itself.

---

**Note:** The article is a part of a Research Project "The meaning of statutory language in light of selected theories from cognitive linguistics", financed by National Science Centre, Poland (2018/31/D/HS5/03922).

---

**Mateusz Zeifert,** University of Silesia

In this article, I would like to present a semantic theory that has barely been noticed by legal scholars, namely the Natural Semantic Metalanguage theory by Anna Wierzbicka. It differs significantly from most other theories about language that have been applied in law. It comes from the realm of theoretical linguistics, lexical semantics and cultural studies, rather than philosophy of language, formal semantics or cognitive psychology. It may not challenge our concept of law, but it addresses two topics that have often been ignored in the legal theoretical literature, yet whose importance in a globalised world is only growing: clarity and the translatability of legal language.

## 2 Natural Semantic Metalanguage

### 2.1 Anna Wierzbicka as a linguist

Natural Semantic Metalanguage (subsequently referred to as 'NSM') is a semantic theory originated by Anna Wierzbicka, now a Professor Emerita in Linguistics at the Australian National University in Canberra (Australia). Wierzbicka was born in 1938 in Warsaw (Poland). She began the career at the Institute of Literary Research of the Polish Academy of Sciences, where she received her PhD and habilitation. In 1972, she emigrated to Australia and started work at the Australian National University in Canberra, where she has spent the rest of her academic career, continuing cooperation with academics from other countries, mostly Poland and Russia (Gladkova/Larina 2018, 500).

Wierzbicka's position in contemporary linguistics is peculiar, though there should be no doubt that she is an extremely influential scholar (Ye/Bromhead 2020, 1).[1] Because of her national and linguistic background, she is particularly renowned in European academia. For instance, "in Russian linguistics, one is unlikely to find another author who is cited as widely and passionately as Wierzbicka" (Gladkova/Larina 2018, 500). However, she is also very difficult to label in theoretical terms.

Her closest affiliation seems to be with Cognitive Linguistics. She made semantics the primary area of her linguistic inquiry as early as in the 1970s, in the era still dominated by Noam Chomsky and his transformative-generative grammar. She had adopted many core ideas of the second wave of cognitive linguistics, such as the notion of prototypes, human cognition-based perspective on meaning,

---

[1] Wierzbicka's h-index of 87, as shown by Google Scholar (05.07.2022), places her among the most prominent linguists still alive, including: George Lakoff – 107, Ray Jackendoff – 83 and Steven Pinker – 99, with Noam Chomsky being outside anybody's reach – 184.

questioning the strict division between semantics and pragmatics, or treating grammar as meaningful, before they were even articulated in the famous writings of Eleanor Rosch, George Lakoff, Charles Fillmore and Ronald Langacker. When Cognitive Linguistics was conceived as a discipline, she was there; she attended the foundational academic conference in Leipzig in 1980 and was published in the very first issue of "Cognitive Linguistics" journal (Goddard 2018, 3–4).

At the same time, Wierzbicka has been notably critical about some aspects of Cognitive Linguists, such as using prototype as a "catch-all" notion (Wierzbicka 1996, 148–167). Unlike most cognitive linguists, she has not sought for inspiration in cognitive psychology, but rather in general philosophy, literary theory, lexicography and anthropology. Most importantly, she has remained a very outspoken advocate of semantic invariants and componential analysis, thought of a very distinctive type. As a result, Wierzbicka is rarely listed among the key figures of Cognitive Linguistics, with her theory typically being labelled as "borderline cognitive" (Brala 2003, 163; see also: Geeraerts 2016, 12). Couple this with rarely seen consequence in developing her own scientific programme and a highly polemical style, and you build a picture of a very unique, independent, and inspiring thinker.

## 2.2 The basics of the NSM theory

In the mid-1960s, a fellow linguist – Andrzej Bogusławski – instilled in Wierzbicka the idea of *alphabetum cognitationum humanarum* – 'the alphabet of human thought' (Bartmiński 2011, 220). It was the "golden dream" of the German philosopher Gottfried Wilhelm Leibniz (1646–1716). Leibniz believed that there must be a set of *indefinibilia*, concepts so basic that they cannot be defined, because otherwise no comprehension would be possible (Wierzbicka 1996, 11). In other words, the fact that we understand anything must be attributed to the existence of some basic set of concepts that are understood intuitively or "in themselves". Later, these were described as "primitive concepts" or "semantic primitives". They were expected to be common for all humans, i.e. universal. In fact, this idea was shared by many great thinkers of XVII century, including Leibniz, Descartes, Arnauld and Pascal (Wierzbicka 1996, 11–13). However, it remained at the level of philosophical speculation, was quickly deemed utopian and eventually abandoned in XVIII century. In 1977, John Lyons expressed the common view that there remain no advocates of the most extreme form of "semantic universalism", namely the idea that "there is a fixed set of semantic components that are universal in that they are lexicalized in all languages" (Lyons 1977, 331–332). However, this is exactly the idea that has been driving Wierzbicka's work since the 1970s.

Wierzbicka took the "golden dream" of XVII century rationalist philosophers as a starting point for modern linguistic research. In her 1972 book, *Semantic Primitives*, she hypothesised a list of 14 universal lexical units called "semantic primitives" (Wierzbicka 2021, 319–320). After moving to Australia, she established a close collaboration with scholars from various regions around the world. The most notable is Cliff Goddard (currently from Griffith University in Queensland, Australia) who actually came up with the name *Natural Semantic Metalanguage* and who has been the second main contributor to the NSM theory (Wierzbicka 2021, 319). Together with international collaboration came contact with various non-European languages. This allowed Wierzbicka to ground the search for linguistic universals in empirical, cross-linguistic studies of many different languages. Quite surprisingly, the comparative research began revealing more and more semantic similarities across languages, causing the list to grow. Today, after more than four decades of research covering approximately thirty different languages, the list has reached 65 elements (Wierzbicka 2021, 320).

**Table 1:** The list of semantic primes (English).

| | |
|---|---|
| I, YOU, SOMEONE, SOMETHING~THING, PEOPLE, BODY | substantives |
| KINDS, PARTS | relational substantives |
| THIS, THE SAME, OTHER~ELSE | determiners |
| ONE, TWO, SOME, ALL, MUCH~MANY, LITTLE~FEW | quantifiers |
| GOOD, BAD | evaluators |
| BIG, SMALL | descriptors |
| KNOW, THINK, WANT, DON'T WANT, FEEL, SEE, HEAR | mental predicates |
| SAY, WORDS, TRUE | speech |
| DO, HAPPEN, MOVE | actions, events, movement |
| BE (SOMEWHERE), THERE IS, BE (SOMEONE/SOMETHING) | location, existence, specification |
| (IS) MINE | possession |
| LIVE, DIE | life and death |
| WHEN~TIME, NOW, BEFORE, AFTER, A LONG TIME, A SHORT TIME, FOR SOME TIME, MOMENT | time |
| WHERE~PLACE, HERE, ABOVE, BELOW, FAR, NEAR, SIDE, INSIDE, TOUCH | place |
| NOT, MAYBE, CAN, BECAUSE, IF | logical concepts |
| VERY, MORE | augmentor, intensifier |
| LIKE | similarity |

The current list of semantic primes is presented in Table 1. For ease of use, they are grouped into various "onto-syntactical" categories. Several caveats should be made here. Although often presented in that way, semantic primes are not, strictly speaking, words (lexemes); "Semantic primes exist, not at the level of whole lexemes, but as the meanings of lexical units" (Goddard/Peeters 2010, 463). They are often referred to as *word-meanings* and understood as concepts expressed in language in the form of either separate words, bounded morphemes, or fixed phrases (phrasemes) (Wierzbicka 2021, 327). They may have combinatorial variants used in various grammatical contexts (known as *allolexes*), i.e. *I/me* in English. On the other hand they can have polysemic extensions, in which case they refer to only one of several meanings of a particular word. For instance, the Polish exponent of the semantic prime *feel* is *czuć*. However, in Polish *czuć* is a polysemous word and can also be translated as 'to smell' (Wierzbicka 1996, 25–28).

With over 7,000 languages spoken around the world, proving that some element is lexicalised in *every* language is virtually an impossible task. The universality claim of NSM is therefore a hypothesis that is yet to be fully confirmed. So far it has been tested on approximately thirty languages from all parts of the world and from diverse language families.[2] Granted, it amounts only to a very small sample of all human languages. At the same time, however, it is roughly thirty times more than most other semantic theories ever take into consideration. This is actually a recurrent theme of criticism employed by Wierzbicka against linguists, psychologists and philosophers alike: that they base their theories about language, meaning and thinking on concepts expressed solely in English, without acknowledging the fact that other languages do not have words capable of conveying the same meaning: "[. . .] the conviction that one can understand human cognition, and human psychology in general, on the basis of English alone seems short-sighted, if not downright ethnocentric" (Wierzbicka 1997, 8; Levisen 2018).

Semantic primes are claimed to be not only *universal*, but also *primary* (or *primitive*), meaning that they are "so simple that they cannot be further explained or defined. They are analogous to chemical elements, which cannot be broken down into any other elements. A semantic primitive, in principle, is a meaning that

---

[2] Including: English, Russian, French, Spanish, Polish, Danish, Italian, Ewe, Amharic, Arabic, Malay, Japanese, Mandarin Chinese, Cantonese, Korean, Vietnamese, Mbula (PNG), East Cree, Yankunytjatjara, Koromu, and others. Information from the Natural Semantic Metalanguage website: https://intranet.secure.griffith.edu.au/schools-departments/natural-semantic-metalanguage/what-is-nsm (28.07.2021).

resists further explanation or decomposition" (Goddard 2018, 310). Not all universal meanings have to be primary. Concepts such as *man, woman, mother, sun, moon, stars, head, hands,* or *legs* are likely to be lexicalized in all human languages (Goddard/Peeters 2010, 468). However, they can be further analysed into simpler terms. This is why they are not included in the list. The chemistry metaphor is useful once again here: they can be thought of as fixed compounds made of basic elements and called *semantic molecules*. Because of their pervasiveness, they also play an important role in the NSM framework (Goddard/Peeters 2010, 467–469).

Semantic primes constitute the lexicon of Natural Semantic Metalanguage. It is complemented by the "universal grammar", i.e. the combinatorial (syntactical) properties of semantic primes (Goddard/Peeters 2010, 473). These involve information that primes of a certain category can combine with primes from another category (i.e. substantives with specifiers), as well as syntactical frames for predicate primes that specify their valency options. Formal realisations of sentences in different languages may vary (i.e. word order in a sentence), but the underlying combinatorial properties remain unchanged (Goddard/Peeters 2010, 473).

Semantic primes, together with universal grammar, form the Natural Semantic Metalanguage. This is a mini-language, a carefully crafted subset of all natural languages that provides a unique a tool for semantic representation. It is mostly used to formulate semantic explications. This is a sort of reductive paraphrase – an attempt to say the same thing, but using only semantic primes and their universal grammar. The main feature of NSM is its translatability. Paraphrases formulated in NSM can be expressed in any of the thirty-some languages currently "supported" by NSM theory, without any distortion in meaning. In other words, once we "break down" a word, sentence or idea into NSM explication, its meaning can be grasped by anybody, regardless of linguistic and cultural differences.

## 2.3 Some examples

Let us now turn to some examples. First is the meaning of the English verb *to kill*. It is a classic example used by generations of linguists and most famously defined as 'cause to die' or 'cause to become not alive'. These definitions, as intuitive and trivial as they may seem, have been subjected to convincing critique (Wierzbicka 1975, 491–492). Some of their deficiencies should be clear for anybody with a legal background. For instance, they do not account for the difference between direct and indirect causation. If Peter leaves his child in his car on a hot summer day, there is a risk that he will 'cause her to die', but we would not normally say that he will 'kill' her. The following NSM explication of *to kill* seeks to amend those shortcomings (Goddard/Peeters 2010, 465):

*Someone X killed someone Y* =
someone X did something to someone else Y
because of this, something happened to Y at the same time
because of this, something happened to Y's body
because of this, after this Y was not living anymore

First, notice that a NSM explication involves paraphrasing not a single, isolated word, but rather the whole sentence. This is an important feature that helps overcome the problem of polysemy. Next, the meaning of the word in question (here: *to kill*) is paraphrased in a series of sentences using only semantic primes and their universal grammar. Note how this paraphrase avoids the problems mentioned above. Instead of a simple 'cause – effect' structure, it describes several steps that include an action by the agent X with an immediate effect on the patient Y, followed by a change in Y's body, followed by Y's death.

Secondly, consider two English adjectives: *sad* and *unhappy* (Goddard/Peeters 2010, 466):

*X felt sad* =
someone X felt something bad
someone can feel something like this when this someone thinks like this:
    "I know that something bad happened
    I don't want things like this to happen
    I can't think like this: I will do something because of it now
    I know that I can't do anything"

This paraphrase depicts a prototypical cognitive scenario that serves as a reference situation for the reader. It is a typical NSM strategy for dealing with words expressing emotional and volitional states that are notoriously difficult to define. Obviously, *sad* is an indeterminate concept in the sense that there are an unlimited number of reasons for, as well as symptoms of, being *sad*. NSM deals with this indeterminacy by focusing instead exclusively on the mental state of the subject and employing a subjective (i.e. human) as opposed to objective, perspective to define the term.

Now, consider the explication of the word *unhappy*.

*X felt unhappy* =
someone X felt something bad
someone can feel something like this when this someone thinks like this for some time:

> "some very bad things happened to me
> I wanted things like this not to happen to me
> I can't not think about it"
> this someone felt something like this
> because this someone thought like this

*Sad* and *unhappy* have very similar meanings. However, *unhappy* includes a stronger negative evaluation. We may say "I feel a little sad", but not: "I feel a little unhappy". In addition, *unhappy* is more personal. We may (and even should) be sad because of the Russian aggression against Ukraine, but we are normally unhappy only because of things that have happened to us. The differences between the NSM explications of both words articulate these subtle nuances of meaning, which are often lost in more traditional definitions. At the same time, they both successfully avoid the vicious circles (defining *ignotum per ignotum*) that often plague lexicographical works (Goddard 2018, 307).

## 2.4 Escaping theoretical dichotomies

As the above explications show, NSM's approach to semantic analysis is quite original. It escapes the popular dichotomies often used to classify linguistic theories. First, it is both formal and natural. It is formal because it employs well-specified and explicit vocabulary and grammar, i.e. a formal system of notation. At the same time, it is natural. Semantic primes are not artificial symbols without meanings, but concepts carved out of natural language. Linguists and philosophers often invent sophisticated, highly technical metalanguage for their analyses, for instance:

1) Bachelor → N, N1, . , N,; (Physical Object), (Living), (Human), (Adult), (Male), (Never-married); < SR >. (Katz 1964, 743)

2) $\text{TO X} = \begin{pmatrix} +b, -i \\ \text{DIM 1d DIR} \\ _{\text{Space}}\text{BDBY}^+ \left( \left[ _{\text{Thing/Space}} X \right] \right) \end{pmatrix}$

(Jackendoff 1991, 36)

The problem is that such technical descriptions still need to be "translated" into natural language in order to be understood by anyone except for their creators. As put by Lyons: "any formalization is parasitic upon the ordinary everyday use of language in that it must be understood, intuitively, on the basis of ordinary language" (Lyons 1977, 12). Substituting words with symbols, spelling them in capital letters or putting in brackets do not automatically make meanings any more specified or intelligible. Apart from that, it is doubtful whether it does a good job of explaining ordinary language concepts if lay people cannot make any sense of it (Goddard 2018,

308). Instead, NSM employs a carefully selected set of ordinary words as a system of notation. That is why NSM explications, like those provided above, do not require much description to be intelligible. They are more or less self-explanatory.

Secondly, NSM escapes the philosophical opposition between linguistic rationalism (i.e. like Chomsky) and empiricism (i.e. like Lakoff). In a sense it is rationalist, even idealist or "platonic" (Geeraerts 2016, 6), as it aims at reconstructing the human conceptual system through an explicit *lingua mentalis*. At the same time, it is empiricist because the *lingua mentalis* is not a result of *a priori* considerations, but of comparative, cross-linguistic studies of different languages (Goddard 2018, 316).

Thirdly, NSM completely breaks apart the dichotomy between linguistic universalism and relativism. More precisely, it is both radically universalist and radically relativist (Goddard 2018, 320). The basic premise of NSM is that all human languages have a common core of lexically expressed meanings. It means that each and every semantic prime has a representative in any given human language. As put by Goddard: "This is certainly the strongest claim about universally lexicalised meanings to be found in the contemporary linguistic literature" (Goddard 2001, 3). On the other hand, Wierzbicka is heavily inspired by the ideas of Humboldt, Boas, Sapir and Whorf, to mention only the most prominent theorists from the "relativist camp". Much of her scientific work can actually be described as "neo-Whorfian" (Goddard 2003, 48). She has always seen language and culture as closely connected and mutually affecting one another. She initiated a whole new discipline of *ethnosynthax* that studies grammar as a vehicle of culture (Wierzbicka 1979). She has put a lot of scientific effort towards warning against the perils of treating English as the "default" language of human thought (Wierzbicka 2014). In other words, the strong claim about linguistic universals does not stop her from investigating the matrix of culture-specific senses, norms and traditions. On the contrary, it is only through the Natural Semantic Metalanguage – the ultimate *tertium comparationis* – that we can recognise and appreciate the infinite diversity of linguistic systems. Paradoxically, "[t]he hypothesis of 'linguistic relativity' makes sense only if it is combined with a well thought-out hypothesis of 'linguistic universality'" (Wierzbicka 1997, 22).

# 3 NSM and the law

## 3.1 Previous applications

NSM theory has so far not drawn much attention from legal scholarship. It is occasionally referred to by authors with a linguistic background. For instance, Peter Tiersma briefly mentions NSM in his discussion about the possibility of drafting statutes in plain language. He admits that it is an interesting approach, but concludes that the length of NSM paraphrases, as compared to normal statutory language, would probably exceed the advantages of clarity it provides (Tiersma 2006, 48–49). Lawrence Solan quotes Wierzbicka on several occasions while discussing some more theoretical aspects of language, namely the opposition between the classical and prototype approach to lexical semantics. He correctly recognises that she holds a somewhat middle position – claiming that word meanings do have definitions, but these definitions refer to mental states, images and scenarios, rather than objective features of the objects "out there", and thus often lead to prototypical effects (Solan 1998, 70–74; Solan 2001, 257). There are certainly more references to Wierzbicka in the legal-linguistic literature (see: i.e. Bajčić 2017, 115–116; Galdia 2017, 413; Durant 2018, 37; Ainsworth 2018, 266), though they are rarely longer than a few sentences.

Interestingly, some NSM scholars have published papers on legal matters. Cliff Goddard contributed to the ongoing debate on statutory interpretation with his 1996 article *Can linguists help judges know what they mean? Linguistic semantics in the court-room* (Goddard 1996). The paper addresses the controversy about the use of dictionary definitions by judges and discusses some potential alternatives that are especially appealing to linguists as expert witnesses. It also introduces the lawyer-reader to the NSM theory and provides NSM explications of several legally-significant concepts, such as *enterprise, reckless,* and *sudden*. The overall conclusion by Goddard is pessimistic, however. He states that the usefulness of linguists (and linguistics) as experts on meaning in legal practice is quite limited, both because of the specific role of ordinary meaning in law application and the underdevelopment of semantics as a scientific discipline (Goddard 1996, 269–270).

Ian Langford, a student of Wierzbicka, published an article on the semantics of selected crimes (Langford 2000) and later wrote a doctoral dissertation entitled *The semantics of crime: a linguistic analysis* (Langford 2002). His main idea was to "add to our knowledge about the semantics of crime in English by analysing the meaning of expressions referring to crimes in both ordinary and legal language" (Langford 2002, 3). He uses corpus research to discover the ordinary meaning of words such as *murder, rape, robbery, hijack* and *assault*, as well as statutory definitions or legal textbooks to establish their legal meaning. Then he formulates NSM explications of their meanings. Finally, he proposes several forensic applications of NSM, including: court interpreting and translating, formulating statutory definitions, police cautions, and jury instructions (Langford 2002, 337–367).

Finally, Anna Wierzbicka herself wrote a paper *'Reasonable man' and 'reasonable doubt': the English language, Anglo culture and Anglo-American law* (Wierzbicka 2003). This is a very interesting, albeit purely linguistic, discussion about two fundamental concepts of Anglo-American law. In her typical, diachronic analysis, full of literary examples, Wierzbicka reconstructs a surprising historical change in meaning. She also proposes NSM explications for *reasonable man* and *reasonable doubt*.

Altogether, NSM's potential has not been fully utilised in law. On the one hand, legal scholars tend to view it as a curiosity with no direct application in the legal domain. On the other hand, the contribution from NSM scholars was so far, all things considered, fairly insignificant. Additionally, in their analyses, NSM scholars naturally adopt a linguistic perspective that a lawyer or jurist may find slightly naïve and unrealistic.[3] Still, I believe that there is much more for legal scholars to be learned from Anna Wierzbicka and her colleagues. I will now proceed to discuss several possible applications of NSM theory in a legal context.

## 3.2 Comprehensibility of legal texts

One obvious area of the application of NSM is the comprehensibility of legal texts. Legal texts are notorious for being incomprehensible for laypeople. The complains about legal language being obscure and complicated go back to at least the sixteenth century, when it was mocked by intellectuals such as Sir Thomas More and Jonathan Swift (Langford 2002, 15–20). In 1963, David Mellinkoff published his seminal book *The Language of the Law* (Mellinkoff 2004), starting a new discipline of legal linguistics. Arguably the most influential part of the book is the scathing critique of contemporary Legal English. Mellinkoff identified nine main characteristics of the genre contributing to its "uncommon touch", most of which concern vocabulary (Mellinkoff 2004, 11–23). Over the years, more scholars turned their attention to legal language and broadened our understanding of its distinctive features, their origins and functions. For instance, in his influential 1999 book *Legal Language*, Peter Tiersma paid a lot of attention to formal (i.e. grammatical and stylistic) aspects of legal documents (Tiersma 1999, 51–86).

The critique of legal language, expressed in the academic world by Mellinkoff, Tiersma, and many others, also took a more practical form. The 1970s saw the emergence of the plain language movement, starting in the banking sector but quickly expanding into legal spheres. It has been especially influential in English speaking countries. Many governments have adopted drafting guidelines for administrative agencies, such as the U.S. Plain Writing Act of 2010 or the Australian Plain English Manual of 1993. The idea of plain legal language is to enhance the comprehensibility

---

**3** Consider, for instance, Langford's proposal that "[. . .] in writing a criminal code, the conceptual structure can take the ordinary meanings as a starting point and as it were, 'gloss' the extra legal components of meaning on to the ordinary meaning" (Langford 2002, 331), which assumes that the "ordinary meaning" of legal terms is commonly known and uncontroversial.

of legal documents: "Plain language has to do with clear and effective communication – nothing more or less" (Kimble 1994, 52). There are countless official documents, guidelines, booklets, manuals and scholarly papers that propose ways and means of achieving this goal, addressing various aspects of written communication.

There are some striking similarities between plain language and NSM theory. Plain language, just like NSM, focuses on the clarity and intelligibility of linguistic expressions. According to Butt, "it is language that communicates directly with the audience for which it is written. It allows the reader to understand on a first reading. It is organised in a way that meets the reader's needs, not the writer's needs" (Butt 2012, 28). NSM seeks to achieve this through the use of a very limited subset of natural language, namely semantic primes. Plain language advocates, quite similarly, suggest avoiding technical, archaic, formal, foreign or otherwise unusual vocabulary (Garner 2001, 62). Plain language shares the founding idea of NSM that complex ideas can be expressed through simple linguistic forms: "Plain language may not be able to simplify concepts, but it can simplify the way concepts are expressed. Used properly, plain language clarifies complex concepts" (Butt 2012, 30). Plain language also promises to deliver clarity without sacrificing precision: "Plain language is usually *more* precise than traditional legal style. The imprecisions of legalese are just harder to spot" (Kimble 1999, 50). The transparency provided by plain language techniques make it easier to identify and deal with possible deficiencies: "plain language helps expose errors. In contrast, legalese tends to hide inconsistencies and ambiguities, because errors are harder to find in dense, convoluted prose" (Butt 2012, 32). A very similar notion is expressed by Wierzbicka with a reference to NSM paraphrases: "[w]hen a formula (. . .) is found wanting from a legal point of view, its inadequacies can be clearly identified, and an improved formula can be devised" (Wierzbicka 2003, 21).

Apparently, there are many common points between the general goals and assumptions of NSM and plain language, and this fact has actually been acknowledged by some NSM scholars (Goddard/Wierzbicka 2015, 2). When we look into details, however, we note significant differences. First and foremost, advocates of plain language usually refrain from direct appeal to linguistic theories. They ground their advice mostly in common sense and anecdotal examples, rather than in empirical research or statistical data (Assy 2011, 377–380). This approach is radically different from that of NSM scholars, who rely on extensive, cross-linguistic studies. Secondly, plain language methods seem to be targeted at structure, style, grammar and even graphic design, rather than at the vocabulary used in legal writing. It is actually presented as one of its virtues (Kimble 1996, 2). Granted, plain language is a diverse enterprise and there is no established canon of plain drafting principles, but this tendency becomes clear after consulting some of the most influential works by plain language advocates (Schiess 2003, 71–75).

Recommendations addressing the problem of vocabulary are less common. They are either very general, such as: "cut unnecessary words", "prefer shorter words to long ones, simple to fancy"; "use familiar words"; "do not use jargon" (Schiess 2003, 71–74) or very specific, such as: "use must instead of shall" (Schiess 2003, 74); "avoid doubles and triples" (Garner 2001, 67). For instance, in a very influential textbook by Bryan A. Garner, in a chapter entitled "Choosing your words", only four out of nine listed principles actually address vocabulary directly. The rest concern grammar and reference, i.e. "turn *–ion* words into verbs" or "refer to people and companies by names" (Garner 2001, 62–84). Plain language advocates suggest using "simple", "familiar" or "strong" words, but rarely – if ever – take the effort to explain what counts as "simple", "familiar" or "strong". At the same time, "[w]hat is impressionistically 'plain' in English isn't necessarily either simple or universal (Goddard/Wierzbicka 2015, 2). Consider the following definitions of *reasonable man/reasonable person* – a fundamental concept of Anglo-American legal culture:

1) 'a fictional person with an ordinary degree of reason, prudence, care, foresight, or intelligence whose conduct, conclusion, or expectation in relation to a particular circumstance or fact is used as an objective standard by which to measure or determine something (as the existence of negligence)' (Merriam-Webster.com Legal Dictionary, https://www.merriam-webster.com/legal/reasonable%20person (28.07.2022)).

2) 'A legal standard used in negligence (personal injury) cases. The hypothetical reasonable person behaves in a way that is legally appropriate. Those who do not meet this standard – that is, they do not behave at least as a reasonable person would – are considered negligent and may be held liable for damages caused by their actions' (Nolo's Plain-English Law Dictionary, Legal Information Institute (https://www.nolo.com/dictionary/reasonable-person-term.html (28.07.2022)).

The first definition is a standard definition from a popular online legal dictionary. The second definition is a plain-English legal definition from a commercial plain English dictionary. When we compare these two definitions, we may notice that the plain version uses simpler vocabulary and simpler grammatical constructions. It avoids the vicious circle of defining *reasonably* using the word *reason*. It reduces the number of difficult terms used to explicate the meaning of *reasonable man*, such as *prudence, care, foresight* and *intelligence*. It avoids complex phrases such as *in relation to* or *by which to measure,* as well as enumerations, such as *conduct, conclusion, or expectation* or *measure or determine.* However, the definition does little to explain what *reasonable person* actually means. It is certainly not enough to define *reasonable person* as someone who 'behaves in a way that is legally appropriate', because the reasonable person standard is usually used to determine what is legally appropriate. It also goes without saying that the reasonable person standard is not only used in personal injury cases, but has a much more universal significance. Lastly, some of the vocabulary used in the definition can hardly be described as "plain", for instance: *negligence, hypothetical, appropriate, considered*. Now, compare this approach to the NSM explication of *reasonable man* provided by Wierzbicka herself (Wierzbicka 2003, 6):

> I think that X is a reasonable man. =
>> I think about X like this:
>>> X can think well about many things
>>> When something happens to X, X can think well about it
>>> Because of this, X can think about it like this:
>>>> 'I know what is a good thing to do now'
>>>> 'I know what is a good thing not to do now"
>>> If other people think about it for some time they can think the same
>>> When I think about X like this, I think: this is good
>>> I don't want to say more
>>> I don't want to say that X is not like many other people

We can see that the latter is a very different approach to defining *reasonable person*. It avoids circularity, namely it does not use *reason* to define *reasonable*. It does not introduce other concepts of similar complexity, like *prudence*. It uses only a few semantic primes that are intuitively understood. In addition, it does not attempt to define *reasonable person* as an abstract notion, but instead it takes human cognition as a point of departure and depicts a prototypical scenario of what a person may think. In addition, it explicitly introduces several important elements that were left out in the plain language definition: It states that the mental capacity of a *reasonable man* is not unlimited. It states that *reasonableness* relates to practical everyday experience rather than abstract speculations and calculations. It states that a *reasonable person* is not an extraordinary one, and so on.

Overall, it seems that NSM can offer substantial support for the idea of drafting clear and comprehensible legal documents. It shares the basic ideas of plain language, but is much more methodologically robust and is based on years of empirical research. In addition, NSM concentrates on vocabulary, which seems to be a weak point of most plain language guidelines and the plain language movement in general.

## 3.3 Semantic analyses

Another potential area of application is legal semantics. Semantic analyses are indispensable both for legal theory and legal practice. Statutory interpretation, doctrinal analyses and the drafting of a legal text – they all include semantic considerations.

Jurist and linguist Lawrence Solan has noticed that "[m]ost battles over legal interpretation are battles about meanings of words" (Solan 2001, 244). However, there are no standardised tools for legal semantic analyses. They are usually carried out using a mix of intuition, dictionary definitions and specific legal vocabulary, including technical terms and foreign (i.e. Latin) words. This poses several problems. Semantic analyses found in legal books, judicial decisions and dictionaries are often circular, indeterminate and unintelligible. As noted by Goddard, "the tradition in lexicography and law alike [is] to eschew simple language in favour of more complex and learned vocabulary"(Goddard 1996, 265).

The very purpose of NSM was to provide a novel and adequate tool for semantic analyses: "The NSM approach can be viewed as a principled and linguistically sophisticated development of traditional ideas about verbal definition" (Goddard 1996, 258). However, its basic assumption differs greatly from many other approaches to lexical semantics. It provides a way of expressing subtleties of meaning using simpler, not more complex, vocabulary. The previous explication of *reasonable man* serves as an example. Now I would like to provide another one. For this I have chosen a concept from criminal law, namely *recklessness*. It refers to the mental (or subjective) element of crime. It is one of the forms of culpability or types of *mens rea*. Other types of *mens rea* in Anglo-American law usually include *intention* and *negligence*. These are interesting concepts because they refer to mental processes that are subjective and notoriously difficult to define. Mental elements of crime are traditionally divided into cognitive (intellectual) and volitional (attitudinal) part (Blomsma/Roef 2019, 179; Duff 2019, 5). There is no "objective" reality to describe, only desires, wants and beliefs. At the same time, I believe this makes them particularly suitable objects of NSM paraphrases.

Recklessness is a form of culpability characteristic for most Common law countries. It constitutes a middle ground between intent and negligence and may be preliminarily defined as 'the conscious taking of an unreasonable risk' (Blomsm/Roef 2019, 189–190). However, its definitions may vary between different jurisdictions and even between different lines of judgments in one jurisdiction. My analyses here are based directly on two English cases widely discussed in the literature on the subject: *Caldwell*[4] and *R v G*[5] (Blomsma/Roef 2019, 191–192). The concept of recklessness in English law has been changing over the years. The 1982 R v *Caldwell* case overruled the previous interpretation of recklessness and established an objective test for recklessness. It has been criticised for blurring the distinction between recklessness and negligence, and hence many other Common law countries have

---

4 R v Caldwell (1981) 1 All ER 961.
5 R v G and another (2003) UKHL 50.

rejected it (Langford 2002, 133). The decision was overruled in the 2003 *R v G and another* case that once again established the subjective test. The differences between these two accounts of recklessness are quite subtle. They revolve around the notions of awareness, foresight, obviousness, risk, (un)reasonableness, etc. Below, I propose NSM explications of these two legal meanings of recklessness, which instead are formulated using only a handful of semantic primes, such as *think, know, want*, etc.

*Caldwell* recklessness:

"A person is reckless as to whether property is destroyed or damaged where:
1) he does an act which in fact creates an obvious risk that property will be destroyed or damaged and
2) when he does the act he either has not given any thought to the possibility of there being any such risk or has recognised that there was some risk involved and has nonetheless gone on to do it."

In other, slightly more abstract, words: "A person acts recklessly when he either realises there is a risk and takes it anyway, or when he fails to see a risk that, by the objective standard of a reasonable man, he ought to have seen"(Blomsma/Roef 2019, 191).[6]

This definition may be turned into the following NSM explication:

X does something recklessly =
X does something
Something bad may happen because of it
    a) It may be like this:
        X knows that something bad may happen because of it
        When X does it, X does not want to think about it
    b) It may be like this:
        X does not know that something bad may happen because of it because he does not want to think about it
        When other people think about it they will think: "something bad may happen because of it"
When other people think about it they will think: "it is not a good thing to do now"
X may know that other people will think this way.

---

[6] Consider also the original passage: "[A] person charged with an offence [. . .] is reckless as to whether or not any property would be destroyed or damaged if (1) he does an act which in fact creates an obvious risk that property will be destroyed or damaged and (2) when he does the act he either has not given any thought to the possibility of there being any such risk or has recognised that there was some risk involved and has none the less gone on to do it."

The explication is divided into two sections, because of the distinction in the cognitive requirement introduced in *Caldwell*. Section (a) paraphrases the typical situation in which a person is aware of the risk: "he [. . .] realises there is a risk and takes it anyway." In NSM words it can be paraphrased as "X knows that something bad may happen because of it." The line "When X does it, X does not want to think about it" serves to delineate reckless risk-taking from intentional risk-taking, which, conversely, should be interpreted as a kind of intent, i.e. a different form of culpability. It also expresses the notion of disregarding the risk, which in the above definitions is expressed indirectly by the words *nonetheless* and *anyway*.[7] Section (b) paraphrases the alternative situation in which a person is not aware of the risk he should have been aware of: "fails to see a risk that [. . .] he ought to have seen." I have chosen the line "he does not want to think", rather than simple "he does not think", because it better captures the blameworthiness of not thinking about the consequences expressed by the words *fails to see* and emphasises the fact that he could have known about the consequences if he had "given any thought to the possibility of there being such a risk." The reasonable person standard is captured by the reference to what people will think about the whole situation. Alternatively, we could, of course, make reference to the NSM explication of reasonable man provided by Wierzbicka. The last two lines, common for both sections, refer to the requirement that the risk taken by the offender be obvious, unreasonable or unjustifiable. Here, again, the concept of *other people* serves as a proxy for the standard of reasonable person (Blomsma/Roef 2019, 192).

Now, compare this with the alternative *R v G* recklessness:

"A person acts recklessly [. . .] with respect to –
i. A circumstance when he is aware of a risk that it exists or will exists;
ii. A result when he is aware of a risk that it will occur;
And it is, in the circumstances known to him, unreasonable to take that risk."

NSM paraphrase:

X does something recklessly =
X does something
Something bad may happen because of it
X knows that something bad may happen because of it

---

[7] For reference, consider the definition of recklessness from the American Model Penal Code: "A person acts recklessly with respect to a material element of an offense when he consciously disregards a substantial and unjustifiable risk that the material element exists or will result from his conduct. The risk must be of such a nature and degree that, considering the nature and purpose of the actor's conduct and the circumstances known to him, its disregard involves a gross deviation from the standard of conduct that a law-abiding person would observe in the actor's situation" (Section 2.02, 2c).

> When X does it, X does not want to think about it
> When other people think about it they will think: "it is not a good thing to do now"
> X may know that other people will think this way.

This explication is shorter, as it only includes section (a) from the previous definition. For the sake of simplicity, the distinction between recklessness with respect to a circumstance and result is ignored here, as in many doctrinal definitions. The subjective standard of recklessness requires that the offender is aware (*knows*) of the risk created by his actions. However, the awareness requirement does not relate to the unreasonableness of the risk. This is captured by the last two lines introducing the "objective" criterion: "When other people think about it . . ." Alternatively, this could be substituted with a reference to the explication of the concept of *reasonable man*.

As you can see, these paraphrases use only the very restricted NSM vocabulary and simple grammatical constructions. They avoid complex vocabulary traditionally used in this context, such as *awareness, (un)reasonable, realisation, disregard, circumstance*, etc. Yet, they arguably succeed in explaining the semantic complexity of the respective concepts and neatly present the subtle differences between them. The cognitive (intellectual) element (awareness) is expressed by the universal word *know*. The volitional element is expressed by the universal word *want*. In recklessness, the volitional element is barely present (Blomsma/Roef 2019, 190), hence it is presented in a negative form: "does not want to think." Paraphrases such as these are short, unambiguous, precise, simple and self-explanatory. They can be used for various legal and forensic purposes, including doctrinal analyses, judicial interpretation of statutes and precedents, writing police warnings and jury instructions (Langford 2002, 337–367). They can also be used in comparative law as a means for comparing concepts from different languages and different legal systems, as will be discussed in the next section.

## 3.4 Comparative law

One of the main methodological issues in comparative law is the problem of a common comparative denominator or *tertium comparationis* (van Reenen 1995, 176; Brand 2007, 409–459). Every comparison of two or more different legal traditions, systems, or institutions presupposes some common ground between them that makes the comparison possible (Hoecke 2015, 27). *Tertium comparationis* forms the conceptual apparatus with which the comparatist approaches his discipline and "provides (or fails to provide) him with the key to access the positive legal reality" (van Reenen 1995, 198). Comparative legal scholars continue to debate on the nature, role and very existence of *tertium comparationis*. Some see it in the function of legal institutions: "incomparables cannot usefully be compared, and in law the only things that are comparable are those which fulfil the same function" (Zweigert/Kötz 1998, 34; Michaels 2006, 367). Some seek it in the supranational ideal, "higher", or natural law, Gustav Radbruch's *richtige Recht* (van Reenen 1995, 177). Others claim that it can only be found in objective social reality (van Reenen 1995, 184). Still others propose a certain philosophical concept of law to serve as the *tertium comparationis* (van Reenen 1995, 197). Moreover, some comparative legal scholars deprecate the very idea of *tertium comparationis* as misleading, arguing that there can be no such thing as a neutral referent (Frankenberg 1985, 415).

Mark van Hoecke has expressed the opinion that the continued search for *tertium comparationis* in comparative law, understood as some external, neutral, objective element, is misguided. What is really needed is creating a language capable of describing concepts from different legal cultures in a relatively neutral way: "[i]nstead of looking for *tertia comparationis*, legal comparatists should, indeed, through their research, develop such a comparative second-order language" (Hoecke 2015, 28).[8] In a roughly similar vein, Oliver Brand has argued for a conceptual approach in comparative law methodology. In his theory, the role of *tertium comparationis* is played by concepts which meet certain criteria of neutrality, unambiguity, and context-independence (Brand 2007, 440). He hopes that this method, over time, "will establish a common reference system in the form of the concepts that it develops" (Brand 2007, 463).

It seems that NSM is perfectly suited to fill the roles sketched out above, and much more. It is a second-order language (i.e. metalanguage) designed to analyse and compare concepts from various languages and cultures – "an invaluable descriptive tool for the analysis and contrastive study of meaning-related phenomena in all languages: a *tertium comparationis* for cross-linguistic study and language typology" (Goddard/Peeters 2010, 460). It is also truly universal, based on cross-linguistic research and explicitly targeted against ethnocentrism in any form. As a side note, it is worth mentioning that both comparative law and NSM claim to have the same spiritual father in German philosopher Gottfried Wilhelm Leibniz (Frankenberg 1985, 427; Eser 1997, 495).

Let us continue with the topic of the subjective element of crimes. Different national (and international) bodies of law distinguish close, but not identical types of *mens rea*. The terminology includes numerous English and Latin terms (not to mention terms from other languages), which are similar, but never quite

---

[8] Van Hoecke is actually quite sceptical as to the possibility of working out the really universal metalanguage and he explicitly contents himself with a second-order language of merely bilateral validity (Hoecke 2015, 28).

equivalent: *direct intent, indirect intent, oblique intent, conditional intent, dolus, dolus directus, dolus indirectus, dolus eventualis, recklessness, negligence, conscious negligence, unconscious negligence, culpa*, etc. The vocabulary used to define these terms is also highly problematic, often Latin-laden or metaphorical: "*Mens rea* [. . .] is still one of the most complex areas of criminal law, in most part, because so many imprecise and vague terms are used to define the mental element" (Badar 2013, 16). As a result, comparative legal studies of *mens rea* are inherently risky, as the researcher first needs to overcome conceptual sinking sands and deal with terminological ambiguities. Comparative and international law scholars like to remind that "writing about intent in different jurisdictions and legal systems entails great challenges" (Lekvall/Martinsson 2020, 101). Once again, this makes *mens rea* the perfect testing ground for NSM in a comparative law context.

In the previous section, I provided an NSM explication of the meaning of recklessness. However, recklessness has no direct counterpart in most Civil law jurisdictions. Instead, Roman-influenced legal systems typically adopt a broader notion of intent, encompassing not only direct and indirect intent, but also what is known as conditional intent, often called *dolus eventualis*.[9] The relation between recklessness and *dolus eventualis* is a very popular subject of comparative analyses. There is an ongoing debate about whether they should be viewed as effectively equivalent or distinct (Lekvall/Martinsson 2020, 104). One's opinion is often a matter of perspective: "To a criminal lawyer trained in the civil law it is fairly uncontroversial to consider *dolus eventualis* as a subcategory of intent. [. . .] But to a U.S. criminal lawyer, the idea that *dolus eventualis* is a form of "intent" is nonsensical [. . .] (Ohlin 2013, 83). The difference between the two is by no means purely academic. For instance, in international criminal law it amounts to "nothing less than the distinction between the terrorist and the soldier"(Ohlin 2013, 130).

Most authors agree that *dolus eventualis* covers some, but not all, cases of Anglo-American recklessness (Duff 2019, 6). Both concepts are claimed to serve a similar function, yet focus on slightly different aspects of the offender's mental state (Blomsma/Roef 2019, 189–190). However, their exact relation is difficult to establish. Part of the problem is terminological. As already mentioned, we lack a neutral language to describe legal concepts, even less if we consider concepts not only from various legal cultures, but also encoded in various languages. This problem basically disappears once we reach for NSM paraphrases.

---

**9** A noticeable exception is the criminal law of South Africa, which, although it generally subscribes to the Common Law family, traditionally recognises *dolus eventualis* as a form of intent (Awa 2019, 152–165; Tsuro 2016, 2).

Obviously, neither recklessness nor *dolus eventualis* present crystal-clear, established conceptual categories. As already discussed, recklessness has several definitions. For the following analysis, I will use the explication of *R v G* recklessness as formulated in the previous section.[10] It offers a good approximation of how this concept is currently understood in most Common law jurisdictions:

> X does something recklessly. =
> X does something.
> X knows that something bad may happen because of it
> When X does it, X does not want to think about it
> When other people think about it they will think: "it is not a good thing to do now"
> X may know that other people will think this way.

Analogically, there are numerous interpretations of *dolus eventualis*. I will refer to the definition provided in the Polish Criminal Code from 1997. It generally conforms with the concept of *dolus eventualis* as defined in other civil law countries, such as Germany or Sweden, but it adds a slightly more "exotic" touch and an additional linguistic challenge to the analysis. First, consider the original Polish version:

> Czyn zabroniony popełniony jest umyślnie [*dolus eventualis*], jeżeli sprawca [. . .] przewidując możliwość jego popełnienia, na to się godzi.[11]

Luckily, we have a Polish version of the NSM. Therefore, to avoid the risk of distorting meaning in the translation process, the definition should be turned into a Polish NSM explication:

> *Osoba X robi coś umyślnie [dolus eventualis]* =
> X robi coś
> X wie, że coś złego może stać się z tego powodu
> Kiedy X to robi, X myśli:
>     "Wiem, że to może się stać. Chcę to zrobić"

Now that we have a Polish NSM explication, we can translate it into potentially any other language, because the vocabulary and grammar used in it are allegedly

---

[10] Here, I cut the second line "Something bad may happen because of it" as, outside of the previous context, it is redundant. It merely repeats the information explicitly encoded in the next line "X knows that something bad may happen because of it."

[11] The following direct translation may be offered: 'A prohibited act is committed intentionally, if the perpetrator [. . .] foreseeing the possibility of its commission, agrees to it'. The ending phrase "na to się godzi", translated here as 'agrees to it' has a slightly archaic feel to it. Alternatively, it may be also translated as 'accepts it' or 'reconciles with it'. Both translations conform with the vocabulary used in academic discussions about *dolus eventualis* in other countries, see i.e. (Blomsma/Roef 2019, 187; Lekvall/Martinsson 2020, 103–104).

universal. All translations should be considered semantically equivalent versions of the same explication. Here is the English version:

> Someone X does something <u>intentionally</u> *(dolus eventualis)* =
> X does something
> X knows that something bad may happen because it
> When X does it, X thinks:
>     "I know that it may happen. I want to do it"

Finally, we may compare this explication with that of recklessness. The first two lines of each are identical. The perpetrator does something and he is aware that it may result in committing a crime (here referred to as "something bad"). We may conclude that the cognitive (intellectual) component of both types of *mens rea* is the same. This conforms to the views expressed in the comparative law literature (Blomsma/Roef 2019, 189–190; Duff 2019, 5; Chiesa 2018, 591–592). The difference lies in the volitional (attitudinal) component. Recklessness does not require any particular attitude towards the possibility of committing a crime. On the contrary, *dolus eventualis* requires a certain kind of attitude from the perpetrator. Depending on the adopted theory, it can be described as *acceptance* or *indifference* (Chiesa 2018, 590; Kowalewska 2013, 57–78). I believe that the paraphrase is broad enough to encompass both accounts. Note that the explication of *dolus eventualis* does not share the last two lines of the recklessness explication referring to the objective probability of the risk and unreasonableness of undertaking it. This is due to the fact that the Polish Criminal Code is deliberately silent on this matter (Kowalewska 2013, 65). Similarly, in German law there is no threshold of the probability of risk, unlike in some other civil law jurisdictions (Blomsma/Roef 2019, 183). If needed, in a particular context, those two lines can be added to the formulation.

A full-fledged comparative analysis would naturally require much more, i.e. an explication of the remaining modes of culpability in both legal traditions. However, even such a limited comparison reveals crucial similarities and differences between common law recklessness and civil law *dolus eventualis*. Additionally, it does so in a simple, self-explanatory, easily translatable way. It avoids complex, culture-dependent vocabulary. It is ready to be extended to other languages, due to the universality of NSM formula. It is worth noting that the topic of applying NSM to comparative law in EU context was already hinted by Bajčić in her brilliant 2017 book on legal semantics. She expressed the opinion that the NSM approach in a legal context "would soon hit a wall due to the nature of the law and unique categories of each legal systems" (Bajčić 2017: 115). I believe that the contrary is true. The undisputed uniqueness of conceptual categories from various legal systems (and languages) makes the NSM approach all the more suitable. As the long debate on *tertium comparationis* proves, comparative law craves for universal metalanguage.

## 3.5 Supranational law

This leads us to the next area of possible application of NSM, namely law in a cross-linguistic and cross-cultural environment, most notably supranational law. In this context, translation becomes "one of the central linguistic operations in law" (Galdia 2017, 270). This is a very broad and diverse topic, often discussed from the perspective of legal translation or drafting multilingual EU law (Ainsworth 2014; Biel 2014; Chromá 2014; Prieto Ramos 2014; Šarčević 2015; Bajčić 2017). To maintain the previous focus on criminal law, I will instead discuss it from the perspective of international criminal law.

International criminal law rests on the assumption that some crimes are universal and should therefore be punishable regardless of the nationality of the perpetrator, the content of respective national criminal law, the place of their commission, etc. They are "particularly grave offences of concern to the world community as a whole" (Einarsen 2012, 4). At the same time, there is no universal language to talk about those crimes, to define them in legal acts, to provide communication in the courtroom and so on. As a result, legal translation is indispensable to the successful application of international criminal law. International criminal institutions typically employ whole units of professional translators and heavily rely upon their work: "[v]irtually every aspect of the International Criminal Court's work is dependent upon translation and interpretation" (Swigart 2017, 208).

Translation is arguably even more challenging for international criminal law than for other branches of international law or EU law. There are several reasons for this. Firstly, international criminal institutions, such as the International Criminal Court (ICC), have to deal with an unparalleled number of languages. For instance, ICC has two working languages, six official languages, and a virtually unlimited quantity of communication languages and situation languages. In 2013, it was expected to support a total of 45 different languages (Tomic/ Montoliu 2013, 224–230). Secondly, many of those languages are labelled by linguists as *languages of lesser diffusion*. Such languages are often non-standardised, meaning that they may lack written tradition, dictionaries and linguistic experts, not to mention any existing legal terminology. All this makes translation extremely difficult (Tomic/ Montoliu 2013, 234–237; Swigart 2017, 206–209). Thirdly, international criminal institutions are highly dependent on witness testimony, as they usually lack an autonomous evidence-gathering capacity and they often deal with crimes committed in regions with low rates of literacy (Karton 2008, 36–37). Courtroom interpreting is arguably the most stressful and demanding type of translation. Additionally, it often involves the difficult task of translating emotional states, colloquial vocabulary, euphemisms, ethnic epithets and slurs (Karton 2008, 38; Tomic/Montoliu 2013, 236–237). Lastly, there is the problem of "cultural dissonance" between the institution and its beneficiaries (Kelsall 2010, 1). The basic doctrines and concepts of international criminal law are undoubtedly Western in origin. At the same time, they are mostly applied to non-Western societies and individuals.[12] This phenomenon also has a linguistic dimension. Despite legal and political efforts, English has become the sole dominant language of international criminal law (Swigart 2017, 212–215). This poses the risk of ethnocentrism, i.e. treating English concepts as universal and self-explanatory: "[w]ith English at present being the main lingua franca there is a danger that international criminal justice will continue to see itself through the eyes of that language of law and all the cultural luggage that comes with it" (Bohlander 2014, 513).

NSM, once again, seems like a tailor-made solution. It promises the universal metalanguage that is truly neutral and fully translatable. It is highly sensitive to the cultural aspect of language, well equipped to deal with cultural norms, values and practices (Wierzbicka 1997). Most importantly, it is explicitly anti-Anglocentric (Wierzbicka 2014; Goddard 2018, 315). Wierzbicka insists that the number of culture-specific words in English is much greater than most of its users – even language experts and academics – would ever expect. This includes plain English words such as *male, female, mind, fact, friend, reason, sex, deal* as well as more sophisticated vocabulary, such as: *right, wrong, fairness, evidence, violence, victim, commitment, cooperation, competition, intention, freedom, feasible, reasonable, humane, inhumane, respect, equality, domination, discrimination, degradation*, etc. (Goddard/ Wierzbicka 2015, 11–12). Many of these words do not have direct equivalents even in most European languages, not to mention languages of lesser diffusion that can be heard in the ICC's courtroom. Using them carelessly in cross-linguistic communication poses obvious risks.

Notice that some of these words seem virtually indispensable in the international criminal law context. It appears that many crucial words in the legal English vocabulary are culture-specific and difficult to translate. How can we effectively grant rights to victims of international crimes if we cannot directly translate either *right* or *victim* into the languages spoken by societies that should benefit from them? How can we prosecute sexual violence as a universal crime if neither *sex* nor *violence* have equivalents in the languages of persons who suffered it?[13]

Consider, once again, the mental element of crimes. The problems identified previously in this respect are multiplied in an international context. According to the Rome Statute, international criminal law requires intent and knowledge for

---

[12] So far, all the trials conducted before the ICC have involved a defendant from Africa. This is one of the reasons why the ICC is sometimes accused of neo-colonialism (Benyera 2018, 3–4).
[13] For an interesting proposal of a language-neutral definition of torture in the NSM framework see (Mooney 2018).

criminal liability before the ICC. There is a lot of confusion, however, as to how exactly the concept of intent should be understood. Despite almost twenty years of the ICC's jurisprudence, it is "still rife with ambiguities and inconsistencies" when it comes to the topic of criminal intent (Marchuk 2014, 156). Different legal traditions and different national jurisdictions cherish varied notions of intent. As a result, "judges from different countries serving on international courts and tribunals are probably influenced by how this concept is defined and understood in their respective legal systems" (Lekvall/Martinsson 2020, 108). The judges, fully aware of the controversies surrounding the notion of intent in comparative law, "inadvertently transposed a certain degree of confusion in international criminal law" (Marchuk 2014, 156).

The concept of intent is provided with a statutory definition in Article 30(2) of the Rome Statute:

> [. . .] a person has intent where:
> (a) In relation to conduct, that person means to engage in the conduct;
> (b) In relation to a consequence, that person means to cause that consequence or is aware that it will occur in the ordinary course of events.

As the abundant literature on the subject may attest, this definition has not ruled out all possible interpretive doubts (Lekvall/Martinsson 2020;, Ambos 2003; Marchuk 2014, 134–157; Badar 2013; Singh 2020; Van der Vyver 2004). For instance, it is not clear whether *dolus eventualis* or recklessness are supported by it (Badar 2009, 441–468). In addition, it is far from obvious how the relation between general intent, as defined in Article 30, and specific intent, provided in more details in the Elements of Crimes, should be construed (Ambos 2003, 12–40; Marchuk 2014, 134–156; Van der Vyver 2004, 69–72). According to experts, the Rome Statute's definition of intent is unacceptably loose (Marchuk 2014, 156), uses "ambiguous and psychologically imprecise wording" (Mantovani 2003, 32) and "require[s] further clarification and elaboration" (Ambos 2003, 40).

Another problematic aspect is the translatability of the definition. As already mentioned, the word *intention*, despite its seemingly universal significance for criminal law, is an English-specific word. This makes the concept of intention very far from being universal and self-explanatory. It is doubtful whether the definition does a good job of explaining the term *intent* to people from non-Western cultures. Notice that it uses other complex and arguably English-specific or at last Euro-specific expressions, such as *conduct, circumstance, consequence, ordinary course of event*, etc. Now consider the following NSM paraphrase of the definition of *intent*:

Someone X has intent. =
X does something
(a) When X does it, X thinks: "I want to do it"
(b) Something may happen because of it
When people think about it they will think: "It will happen".
It may be like this: When X does it, X thinks: "I want it to happen"
It may be like this: When X does it, X thinks: "I know that it will happen"

The paraphrase makes it plain to see why the definition draws so much academic discussion and conflicting opinions. Section (a) is rather uncontroversial as it obviously covers cases of direct intent (*dolus directus*) – when the perpetrator consciously aims at committing a crime. Section (b) covers two situations. The first is, once again, uncontroversial. The second, however, bears striking similarity to the previous explications of *dolus eventualis* and recklessness. The main difference between Article 30's intent and *dolus eventualis* is that the latter requires acceptance of the risk (stronger volitional element) without requiring certainty about the consequences (weaker cognitive element).

The above issues are already being discussed in the literature. Some authors have expressed the view that Article 30 does cover *dolus eventualis* and even recklessness: "One could infer from Article 30 [. . .] of the Statute, that the mental element [. . .] comprises not only intent *(dolus)*, but also recklessness *(dolus eventualis)*"(Mantovani 2003, 34). Others insist that the provision does not accommodate lower forms of *mens rea*, such as indirect intent or recklessness, because the definition "clearly indicates that the required standard of occurrence is close to certainty" (Badar/Porro 2017, 318). They point to the difference between *will occur* and *might occur,* which was explicitly referred to by the Preparatory Committee and later by the ICC itself (Badar /Porro 2017, 318–319; Badar 2009, 441–442).

Settling this dispute is not my ambition. Rather, I would like to underline that all those subtle semantic differences between various forms of *mens rea* discussed in the legal literature are successfully captured by the proposed NSM paraphrases. The NSM is obviously simpler than the language of the Rome Statute and jurisprudence, as it uses only a handful of primitive concepts. At the same time, I believe it does not sacrifice precision. Arguably, it is even more precise. For instance, expressions like *virtual certainty, practical certainty, close to certainty, certain unless extraordinary circumstances intervene* are substituted with a simple cognitive scenario which encapsulates both the subjective and objective elements: "When people think about it, they will think: 'It will happen'. When X does it, X thinks: 'I know that it will happen.'"

On top of that, NSM paraphrases are universal. This means that they are not bound to any specific language or legal tradition, such as Common law or Civil law. Instead, they may be freely translated into virtually any language and then discussed in a language-neutral and culture-neutral way. This seems to be particularly important for international criminal law because of its inherently multilingual and multicultural nature. For instance, an African language of lesser diffusion is likely not to have words for *intention, cognition, volition, recklessness, reconciliation, ordinary course of events*, etc., but it is almost guaranteed to have words for *think, want, happen, people*, etc.

# 4 Conclusions

The list of possible legal applications of NSM presented in this article should by no means be read as closed. My intention was to bring the NSM theory to legal theory and to provide some examples of how it can be utilised. There is certainly much more to NSM still to be discovered. There are also, as with the implementation of all scientific theories, some problems to be discussed.

Firstly, I am aware that the NSM paraphrases presented in this article may strike readers as simplistic, naïve, or even "childish" (Goddard/Ye 2014, 11–12). They certainly do not resemble traditional statutory definitions or legal analyses. However, if we resist that first impression and actually apply them in the appropriate legal context, we will be able to judge their suitability. I have tried to show that it is surprisingly high, at least in the presented areas.

Secondly, an obvious objection to NSM is that the paraphrases are too long, difficult to "unpack", and we cannot realistically expect them to substitute legal language. With this I fully agree. Writing statutes or judgments in NSM is undoubtedly a utopian idea that nobody has ever supported. Even NSM scholars advertise it "not as the sole language of communication, but as an auxiliary or supplementary language" (Goddard/Wierzbicka 2015, 1). In law, this may mean formulating explications of key concepts for the purposes of explaining them to laypersons, translating them into other languages, or comparing them with other concepts.

Thirdly, one may argue that NSM is too "ambitious" for legal purposes. After all, do we really need concepts that are primary and universal in our legal analyses? Is it actually beneficial if we submit to the methodological rigour of NSM? Apparently, the proponents of NSM have already recognised this problem. They recently launched the Minimal English project, which can be described as an attempt to take NSM "out of the lab" (Goddard/Wierzbicka 2015, 1). It is an application and extension of NSM, "intended for use by non-specialists, and for a wide and open-ended range of functions" (Goddard/Wierzbicka 2015, 1). It consists of semantic primes, universal semantic molecules and other expressions that are relevant for a particular domain of discourse and easily translatable. It is not a rigid methodology, but rather a flexible system open for future modifications (Goddard/Wierzbicka 2015, 12). The hope of its creators is that it can become "a global minimal lingua franca for the elucidation of ideas and explanation of meanings – not only in scholarship but also in international relations, politics, business, law, ethics, education, and indeed in any context where it is important to explain precisely what one means (Goddard/Wierzbicka 2018, 8).

Finally, there is one more general-theoretical remark to be made. I am aware that the overall tone of the article may feel a little too optimistic. Law has witnessed the "colonisation" of countless linguistic, psychological, sociological and other theories, but they were never able to displace legal problems in their entirety. Obviously, I do not see NSM as a miraculous panacea for all linguistic issues in law. Its utility may in fact be quite limited: "a metalanguage is a tool designed to serve specific ends, and [. . .] as the ends are different, so different tools will be appropriate" (Riemer 2006, 377). For instance, while NSM does a pretty good job with abstract concepts, such as *sad, unhappy, intention, recklessness*, etc., it is gets much more messy with concrete concepts, such as *cup, fruit, tiger*, etc. (Allan 2020; Geeraerts 2009, 119–127). This can be perhaps attributed to the fact that NSM focuses on intensional, rather than extensional, aspects of linguistic meaning (Goddard 2018, 329). It attempts to capture the conceptual knowledge of language users, but does not seem to offer a ready-made explanation of the connection between conceptual knowledge and reality (Geeraerts 2009, 124–125). NSM explications tend to capture the prototype of a conceptual category rather than the full scope of its possible applications (Wierzbicka 1996, 148–169). To use a classic legal example: an NSM explication of *vehicle* is not likely to help us decide whether a bicycle is a vehicle (Hart 1958, 607). My point is, however, that this should not automatically discourage us from making use of NSM theory where suitable. If it can guide us in writing clearer, more culture-neutral and more translatable legal texts, then it is definitely worth pursuing.

# Bibliography

Ainsworth, Janet. 2014. *Lost in Translation? Linguistic Diversity and the Elusive Quest for Plain Meaning in the Law*. In Le Cheng, King Kui Sin & Anne Wagner (eds.), *The Ashgate Handbook of Legal Translation*, Ashgate Publishing, 43–56.

Ainsworth, Janet. 2018. *Law and the Grammar of Judgment*. In Janny H. C. Leung & Alan Durant (eds.), *Meaning and Power in the Language of Law*, Cambridge, Cambridge University Press, 259–276.

Allan, Keith. 2020. *On the semantics of cup*. In Helen Bromhead & Zhengdao Ye (eds.), *Meaning, Live and Culture. In Conversation with Anna Wierzbicka*, Canberra, ANU Press, 441–460.

Ambos, Kai. 2003. *Some Preliminary Reflections on the Mens Rea Requirements of the Crimes of the ICC Statute and of the Elements of Crimes*. In Lal Chand Vohrah, Fausto Pocar, Yvonne Featherstone,

Oliver Fourmy, Michael F. Graham & John Hocking (eds.), *Man's Inhumanity to Man: Essays on International Law in Honour of Antonio Cassese*, The Hague, Brill, doi:10.1163/9789004479098_008, 11–40.

Assy, Rabeea. 2011. *Can the Law Speak Directly to its Subjects? The Limitation of Plain Language*. Journal of Law and Society 38(3), doi:10.1111/j.1467-6478.2011.00549.x, 376–404.

Awa, Linus Tambu. 2019. *The Interpretation and Application of Dolus Eventualis in South African Criminal Law*, University of South Africa.

Badar, Mohamed Elewa. 2013. *The Concept of Mens Rea in International Criminal Law: The Case for a Unified Approach*, Bloomsbury Publishing.

Badar, Mohamed Elewa & Sara Porro. 2017. *Article 30. Mental element*. In Mark Klamberg (ed.), *Commentary on the Law of the International Criminal Court*, Brussels, Torkel Opsahl Academic EPublisher, 314–322.

Bajčić, Martina. 2017. *New Insights into the Semantics of Legal Concepts and the Legal Dictionary. tlrp.17.* Amsterdam/Philadelphia, John Benjamins Publishing Company.

Bartmiński, Jerzy. 2011. *Droga naukowa Anny Wierzbickiej. Od składni polskiej prozy renesansowej do semantyki międzykulturowej*, Teksty Drugie (1–2), 218–238.

Benyera, Everisto. 2018. *Is the International Criminal Court Unfairly Targeting Africa? Lessons for Latin America and the Caribbean States*, Politeia 37, doi:10.25159/0256-8845/2403, 1–30.

Biel, Łucja. 2014. *Lost in the Eurofog: The Textual Fit of Translated Law*, Frankfurt, Peter Lang.

Blomsma, Jeroen &David Roef. 2019. *Chapter VII. Forms and Aspects of Mens Rea*. In Johannes Keiler & David Roef (eds.), *Comparative Concepts of Criminal Law. 3rd Edition*, Cambridge, Intersentia, 177–205.

Bohlander, Michael. 2014. *Language, Culture, Legal Traditions, and International Criminal Justice*, Journal of International Criminal Justice 12(3), doi:10.1093/jicj/mqu034, 491–513.

Brala, Marija M. 2003. *NSM within the cognitive linguistics movement: Bridging some gaps*, Jezikoslovlje 4(2), 161–186.

Brand, Oliver. 2007. *Conceptual Comparisons: Towards a Coherent Methodology of Comparative Legal Studies*, Brooklyn Journal of International Law 32(2), 405–466.

Butt, Peter. 2012. *Legalese versus plain language*, Amicus Curiae (35), doi:10.14296/ac.v2001i35.1332, 28–32.

Chiesa, Luis E. 2018. *Mens Rea in Comparative Perspective*, Marquette Law Review 102(2), 575–603.

Chromá, Marta. 2014. *Making sense in legal translation*, Semiotica 201, doi:10.1515/sem-2014-0018, 121–144.

Duff, Robin, A. 2019. *Two Models of Criminal Fault*, Criminal Law and Philosophy 13, doi:10.2139/ssrn.3420148, 643–665.

Durant, Alan. 2018. *Seeing sense: the complexity of key words that tell us what law is*. In Janny H. C. Leung & Alan Durant (eds.), *Meaning and Power in the Language of Law*, Cambridge, Cambridge University Press, 32–70.

Einarsen, Terje. 2012. *The Concept of Universal Crimes in International Law*, Oslo, Torkel Opsahl Academic EPublisher.

Elewa Badar, Mohamed. 2009. *Dolus Eventualis and the Rome Statute Without It?*, New Criminal Law Review 12(3), doi:10.1525/nclr.2009.12.3.433, 433–467.

Endicott, Timothy A. O. 2008. *Herbert Hart and the Semantic Sting*, Legal Theory 4(3), doi: https://doi.org/10.1017/S1352325200001038, 283–300.

Eser, Albin. 1997. *The Importance of Comparative Legal Research for the Development of Criminal Sciences*. In Roger Blancpain (ed.), *Law in motion: recent developments in civil procedure*, *constitutional, contract, criminal, environmental, family & succession, intellectual property, labour, medical, social security, transport law*, The Hague, Kluwer, 492–517.

Frankenberg, Gunter. 1985. *Critical Comparisons: Re-thinking Comparative Law*, Harvard International Law Journal, vol. 26, 411–455.

Galdia, Marcus. 2017. *Lectures on Legal Linguistics*, Peter Lang Edition.

Garner, Bryan. 2001. *Legal Writing in Plain English. A Text with Exercises*, Chicago/London, University of Chicago Press.

Geeraerts, Dirk. 2009. *Theories of Lexical Semantics*, Oxford University Press, doi:10.1093/acprof:oso/9780198700302.001.0001.

Geeraerts, Dirk. 2016. *Prospects and problems of prototype theory*, Diacronia 4, doi:10.17684/i4A53en, 1–16.

Gladkova, Anna & Tatiana Larina. 2018. *Anna Wierzbicka, Words and the World*, Russian Journal of Linguistics, vol. 22, doi:10.22363/2312-9182-2018-22-3-499-520, 499–520.

Goddard, Cliff. 1996. *Can linguists help judges know what they mean? Linguistic semantics in the court-room*, International Journal of Speech Language and the Law 3(2), 250–267.

Goddard, Cliff. 2001. *Lexico-Semantic Universals: A Critical Overview*, Linguistic Typology 5(1), doi: https://doi.org/10.1515/lity.5.1.1, 1–65.

Goddard, Cliff. 2003. *Whorf meets Wierzbicka: variation and universals in language and thinking*, Language Sciences, vol. 25(4), doi:10.1016/S0388-0001(03)00002-0, 393–432.

Goddard, Cliff. 2018. *Ten Lectures on Natural Semantic Metalanguage: Exploring language, thought and culture using simple, translatable words*, Leiden/Boston, Brill.

Goddard, Cliff &Bert Peeters. 2010. *The Natural Semantic Metalanguage (NSM) approach*. In Bernd Heine & Heiko Narrod (eds.), *The Oxford Handbook of Linguistic Analysis*, Oxford, Oxford University Press, 459–484.

Goddard, Cliff & Anna Wierzbicka. 2015. *What is Minimal English (and How to Use it), Briefing Paper for the "Global English, Minimal English" Symposium (July 2015, ANU, Canberra)*.

Goddard, Cliff &Anna Wierzbicka. 2018. *Minimal English and How It Can Add to Global English*. In Cliff Goddard (ed.), *Minimal English for a Global World*, Palgrave Macmillan, 5–28.

Goddard, Cliff & Zhengdao Ye. 2014. *Exploring "happiness" and "pain" across languages and cultures*, International Journal of Language and Culture 1, doi:10.1075/ijolc.1.2.01god, 131–148.

Hart, Herbert L. A. 1958. *Positivism and the Separation of Law and Morals*, Harvard Law Review 71(4), doi:10.2307/1338225, 593–629.

Hart, Herbert L. A. 2012. *The Concept of Law*, Oxford, Oxford University Press.

Hoecke, Mark Van. 2015. *Methodology of Comparative Legal Research*, Law and Method 12, doi:10.5553/REM/.000010, 1–35.

Jackendoff, Ray. 1991. *Parts and boundaries*, Cognition 41, 9–45.

Karton, Joshua. 2008. *Lost in Translation: International Criminal Tribunals and the Legal Implications of Interpreted Testimony*, Vanderbilt Journal of Transnational Law 41(1), doi:10.2139/ssrn.1511946, 1–54.

Katz, Jerrold, J. 1964. *Semantic Theory and the Meaning of „Good"*, Journal of Philosophy 61(23), doi:10.2307/2023019, 739–766.

Kelsall, Tim. 2010. *International Criminal Justice and Non-Western Cultures*, Oxford Transitional Justice Research Working Paper Series 1, 1–5.

Kimble, Joseph. 1994. *Answering the Critics of Plain Language*, The Scribes Journal of Legal Writing 5, 51–85.

Kimble, Joseph. 1996. *Writing for Dollars, Writing to Please*, The Scribes Journal of Legal Writing 6, 1–38.

Kimble, Joseph. 1999. *The Great Myth That Plain Language Is Not Precise*, Business Law Today 9, 48–51.
Kowalewska, Magdalena. 2013. *Zamiar ewentualny w świetle psychologii*, Poznań.
Langford, Ian. 2000. *Forensic Semantics: the meaning of murder, manslaughter and homicide*, International Journal of Speech Language and the Law 7(1), 72–94.
Langford, Ian. 2002. *The semantics of crime. A linguistic analysis*, Canberra, Australian National University.
Lekvall, Ebba & Dennis Martinsson. 2020. *The Mens Rea Element of Intent in the Context of International Criminal Trials in Sweden*, Scandinavian Studies in Law 66, 100–129.
Levisen, Carsten. 2018. *The Grammar of Violence: Insights from Danish Ethnosyntax and the Wierzbicka-Pinker Debate*, Etnolingwistyka. Problemy Języka i Kultury 30, doi:10.17951/et.2018.30.145, 145–167.
Lyons, John. 1977. *Semantics: Volume 1*, Cambridge University Press.
Mantovani, Ferrando. 2003. *The General Principles of International Criminal Law: The Viewpoint of a National Criminal Lawyer*, Journal of International Criminal Justice 1(1), doi:10.1093/jicj/1.1.26, 26–38.
Marchuk, Iryna. 2014. *The Fundamental Concept of Crime in International Criminal Law. A Comparative Law Analysis*, Berlin/Heidelberg, Springer.
Margolis, Eric &Stephen Laurence. 1999. *Concept and Cognitive Science*. In Eric Margolis & Stephen Laurence (eds.), *Concepts: Core Readings*, MIT Press.
Mellinkoff, David. 2004. *The Language of the Law*, Wipf and Stock Publishers.
Michaels, Ralf. 2006. *The Functional Method of Comparative Law*. In Mathias Reimann & Reinhard Zimmermann (eds.), *The Oxford Handbook of Comparative Law*, Oxford, Oxford University Press, 339–382.
Mooney, Annabelle. 2018. *Torture Laid Bare: Global English and Human Rights*. In Cliff Goddard (ed.), *Minimal English for a Global World. Improved Communication Using Fewer Words*, Palgrave Macmillan, doi:10.1007/978-3-319-62512-6_7, 143–167.
Müller Fonseca, Alexandre. 2018. *Hart and Putnam on Rules and Paradigms: A Reply to Stavropoulos*, International Journal for the Semiotics of Law – Revue internationale de Sémiotique juridique 31(1), doi:10.1007/s11196-017-9526-9, 53–77.
Ohlin, Jens David. 2013. *Targeting and the Concept of Intent*, Michigan Journal of International Law 35(1), 79–130.
Prieto Ramos, Fernando. 2014. *Legal Translation Studies as Interdiscipline: Scope and Evolution*, Meta 59(2), doi:10.7202/1027475ar, 260–277.
Reenen, TP van. 1995. *Major theoretical problems of modern comparative legal methodology (1): The nature and role of the tertium comparationis*, The Comparative and International Law Journal of Southern Africa, 28(2), 175–199.
Riemer, Nick. 2006. *Reductive Paraphrase and Meaning: A Critique of Wierzbickian Semantics*, Linguistics and Philosophy 29, doi:10.1007/s10988-006-0001-4, 347–379.
Šarčević, Susan. 2015. *Language and Culture in EU Law: Multidisciplinary Perspectives*, Ashgate Publishing, Ltd.
Schiess, Wayne. 2003. *What Plain English Really Is*, The Scribes Journal of Legal Writing 9, 43–75.
Singh, Avantika. 2020. *Mens Rea in International Law: Nuremberg to the International Criminal Court*, International Journal of Legal Developments and Allied Issues 6(5), 192–205.
Solan, Lawrence M. 1998. *Law, Language, and Lenity*, William & Mary Law Review 40(1), 57–144.
Solan, Lawrence M. 2001. *Why Laws Work Pretty Well, but Not Great: Words and Rules in Legal Interpretation*, Law & Social Inquiry 26(1), 243–270.

Stavropoulos, Nicos. 2001. *Hart's Semantics*. In Jules Coleman (ed.), *Hart's Postscript: Essays on the Postscript to `the Concept of Law'*, Oxford, Oxford University Press.

Swigart, Leigh. 2017. *Linguistic and Cultural Diversity in International Criminal Justice: Toward Bridging the Divide*, Pacific Law Journal (Sacramento, Calif.) 48, 197–217.

Tiersma, Peter M. 1999. *Legal Language*, Chicago/London, University of Chicago Press.

Tiersma, Peter M. 2006. *Some Myths about Legal Language*, Law, Culture and the Humanities 2, doi:10./191117438721061it-035oa, 29–50.

Tomic, Alex &Ana Montoliu. 2013. *Translation at the International Criminal Court*. In Anabel Borja Albi & Fernando Prieto Ramos (eds.), *Legal Translation in Context. Professional Issues and Prospects*, Peter Lang, 221–242.

Tsuro, Janet Audrey. 2016. *An Alternative Approach to Dolus Eventualis*, University of KwaZulu-Natal.

Van der Vyver, Johan. 2004. *The International Criminal Court And The Concept Of Mens Rea In International Criminal Law*, University of Miami International and Comparative *Law Review* 12(1), 57–149.

Wierzbicka, Anna. 1975. *Why "Kill" Does Not Mean "Cause to Die": The Semantics of Action Sentences*. Foundations of language, vol. 13(4), 491–528.

Wierzbicka, Anna. 1979. *Ethno-Syntax and the Philosophy of Grammar*, Studies in Language. International Journal sponsored by the Foundation "Foundations of Language" 3(3), doi:10.1075/sl.3.3.03wie, 313–383.

Wierzbicka, Anna. 1996. *Semantics: Primes and Universals*, Oxford/New York, Oxford University Press.

Wierzbicka, Anna. 1997. *Understanding Cultures Through Their Key Words*, Oxford/New York, Oxford University Press.

Wierzbicka, Anna. 2003. *"Reasonable man" and "reasonable doubt": the English language, Anglo culture and Anglo-American law*, International Journal of Speech, Language and the Law 10(1), 1–22.

Wierzbicka, Anna. 2014. *Imprisoned in English: the hazards of English as a default language*, Oxford, Oxford University Press.

Wierzbicka, Anna. 2021. *"Semantic Primitives", fifty years later*, Russian Journal of Linguistics 25(2), doi:10.22363/2687-0088-2021-25-2-317-342, 317–342.

Ye, Zhengdao &Helen Bromhead. 2020. *Introduction*. In Helen Bromhead & Zhengdao Ye (eds.), *Meaning, Life and Culture (In conversation with Anna Wierzbicka)*, Canberra, ANU Press, doi:10.2307/j.ctv1d5nm0d.6, 1–10.

Zeifert, Mateusz. 2022. *Rethinking Hart: From Open Texture to Prototype Theory – Analytic Philosophy Meets Cognitive Linguistics*. International Journal for the Semiotics of Law – Revue internationale de Sémiotique juridique 35(2), doi:10.1007/s11196-020-09722-9, 409–430.

Zweigert, Konrad & Hein Kötz. 1998. *An Introduction to Comparative Law*, Oxford University Press.

# Index

Abilities  147–149, 150, 152, 168
Accuracy  92, 114, 115
Acquaintance  55–56, 58
Anna Wierzbicka  174–175, 182
Argumentation, legal  2, 13–14, 31, 32, 41–44,
    47–48, 167
Audience  1, 20–24, 30–32, 33–35, 41,
    59, 184

Clarity  79–80, 174, 182, 184
ClearAct project  81–85, 87
Cliff Goddard  176, 182
Comparative law  190–194, 197
Corpus linguistics  119, 124–125, 127–128, 140
Court interpreting  92, 93, 103, 107, 110, 114–115,
    182, 195

Defence (attorney, counsel)  16, 19, 24, 81, 86, 87
Demonstratives  48, 51
Descriptors  127–129, 138, 176

Epistemic primacy  98, 99
Equivalence  101, 109, 115
Ethos  32–33
Evidence, legal  47–49, 55, 61–62, 68, 71, 74, 84,
    195, 196

Fact-finding, legal  49–50, 51, 61

Indexicality  51, 52–55
Inference  48, 51, 52
International arbitration  68–71, 75, 77
International Criminal Court  195
International Criminal Law  192, 195–199
Interpretation, legal  1, 20–22, 27–28, 31, 42, 44,
    47, 61, 74–75, 92–93, 119, 123–125, 140, 147,
    158–161, 165, 169, 182, 186, 190
Investor-state arbitration  76

Justification, legal  30, 31–32, 43–44, 61, 132, 158

kairos  34

Language for Specific Purposes  121–123

Meaning, legal  2, 132, 182, 188
Meaning-use relation (MUR)  148, 150–153, 170

Natural Semantic Metalanguage  174–181,
    199–200
New York Convention  70
Norms and legal argumentation  13, 22, 24–26,
    30, 43, 122, 159, 168, 169–170

Obscurity of legal language  88
Ostension  48–52, 56, 61

Paradigms of art  13, 15, 17, 22
Peirce, Charles S.  53, 56
Persuasion, legal  34, 35, 36, 43–44
Plain language  80, 88, 182, 183–186
Practice(s), legal  2, 41, 42, 70–71, 73, 74, 118, 120,
    121, 145–147, 149, 153, 157, 160, 165, 168,
    169–170, 182, 186
Pragmatics  2, 6, 51, 73–77, 85–88, 93, 100,
    102, 110, 114, 119, 124, 125, 147–149,
    160, 166
Pseudonymization  83–84

Rhetoric, legal  32, 35, 43
Rule(s), legal  31, 51, 153, 156, 165, 169

Semantic objectivity  125–126, 140
Semantics, legal  186–190, 194
Social perception  127, 134
Sociolinguistics  106, 118, 119, 123–127, 140
Sports law  69

tertium comparationis  181, 190–191
Textual representation  123, 126–127, 131–132, 140
Theory, legal  146, 153, 186, 199
Topoi  30, 35–38, 44
Translation, legal  49, 76, 92–94, 101–103, 114–116, 118, 182, 193, 195, 199–200

Value(s)  3, 12, 19, 24, 30, 32, 35, 40, 43, 131, 160, 161–165, 169, 196
Value-laden words  37, 43

Weighing and balancing  38–40
Witness(es)  13, 17, 50, 53, 58, 68, 71, 74, 92, 104, 106–107, 112, 195

www.ingramcontent.com/pod-product-compliance
Lightning Source LLC
Chambersburg PA
CBHW050525170426
43201CB00013B/2093